QM Library
D1187954

Graduate Texts in Mathematics 215

Springer
New York
Berlin
Heidelberg
Hong Kong
London
Milan
Paris
Tokyo

Graduate Texts in Mathematics

(continued after index)

David M. Goldschmidt

Algebraic Functions and Projective Curves

Springer

David M. Goldschmidt
IDA Center for Communications Research—Princeton
Princeton, NJ 08540-3699, USA
gold@daccr.org

Mathematics Subject Classification (2000): 14H05, 11R42, 11R58

Library of Congress Cataloging-in-Publication Data
Goldschmidt, David M.
Algebraic functions and projective curves / David M. Goldschmidt.
p. cm. — (Graduate texts in mathematics ; 215)
Includes bibliographical references and index.
ISBN 0-387-95432-5 (alk. paper)
1. Algebraic functions. 2. Curves, Algebraic. I. Title. II. Series.
QA341.G58 2002
515.9—dc21 2002016004

ISBN 0-387-95432-5 Printed on acid-free paper.

Printed in the United States of America.

9 8 7 6 5 4 3 2 1 SPIN 10864561

Typesetting: Pages created by the author using a Springer TEX macro package.

www.springer-ny.com

Springer-Verlag New York Berlin Heidelberg
A member of BertelsmannSpringer Science+Business Media GmbH

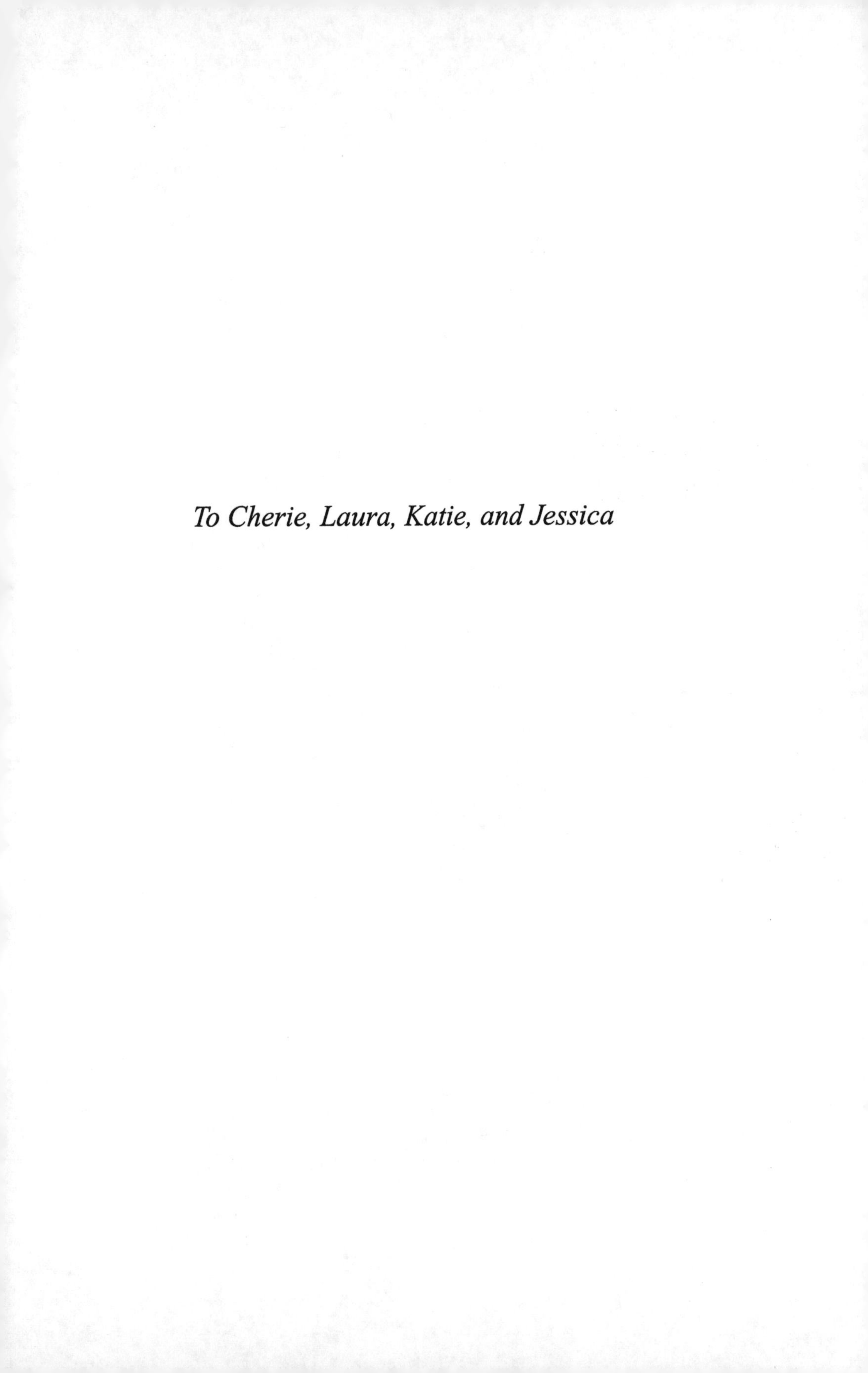

To Cherie, Laura, Katie, and Jessica

Preface

This book grew out of a set of notes for a series of lectures I orginally gave at the Center for Communications Research and then at Princeton University. The motivation was to try to understand the basic facts about algebraic curves without the modern prerequisite machinery of algebraic geometry. Of course, one might well ask if this is a good thing to do. There is no clear answer to this question. In short, we are trading off easier access to the facts against a loss of generality and an impaired understanding of some fundamental ideas. Whether or not this is a useful tradeoff is something you will have to decide for yourself.

One of my objectives was to make the exposition as self-contained as possible. Given the choice between a reference and a proof, I usually chose the latter. Although I worked out many of these arguments myself, I think I can confidently predict that few, if any, of them are novel. I also made an effort to cover some topics that seem to have been somewhat neglected in the expository literature. Among these are Tate's theory of residues, higher derivatives and Weierstrass points in characteristic p, and inseparable residue field extensions. For the treatment of Weierstrass points, as well as a key argument in the proof of the Riemann Hypothesis for finite fields, I followed the fundamental paper by Stöhr–Voloch [19]. In addition to this important source, I often relied on the excellent book by Stichtenoth [17].

It is a pleasure to acknowledge the excellent mathematical environment provided by the Center for Communications Research in which this book was written. In particular, I would like to thank my colleagues Toni Bluher, Brad Brock, Everett Howe, Bruce Jordan, Allan Keeton, David Lieberman, Victor Miller, David Zelinsky, and Mike Zieve for lots of encouragement, many helpful discussions, and many useful pointers to the literature.

Contents

Prefacc vii
Introduction xi

1 Background **1**
1.1 Valuations 1
1.2 Completions 16
1.3 Differential Forms 24
1.4 Residues 30
1.5 Exercises 37

2 Function Fields **40**
2.1 Divisors and Adeles 40
2.2 Weil Differentials 47
2.3 Elliptic Functions 52
2.4 Geometric Function Fields 54
2.5 Residues and Duality 58
2.6 Exercises 64

3 Finite Extensions **68**
3.1 Norm and Conorm 69
3.2 Scalar Extensions 72
3.3 The Different 75
3.4 Singular Prime Divisors 82
3.5 Galois Extensions 89
3.6 Hyperelliptic Functions 93

3.7 Exercises 99

4 Projective Curves **103**
4.1 Projective Varieties 103
4.2 Maps to $\mathbb{P}^n$ 108
4.3 Projective Embeddings 114
4.4 Weierstrass Points 122
4.5 Plane Curves 136
4.6 Exercises 147

5 Zeta Functions **150**
5.1 The Euler Product 151
5.2 The Functional Equation 154
5.3 The Riemann Hypothesis 156
5.4 Exercises 161

A Elementary Field Theory **164**

References **175**

Index **177**

Introduction

What Is a Projective Curve?

Classically, a projective curve is just the set of all solutions to an irreducible homogeneous polynomial equation $f(X_0, X_1, X_2) = 0$ in three variables over the complex numbers, modulo the equivalence relation given by scalar multiplication. It is very safe to say, however, that this answer is deceptively simple, and in fact lies at the tip of an enormous mathematical iceberg.

The size of the iceberg is due to the fact that the subject lies at the intersection of three major fields of mathematics: algebra, analysis, and geometry. The origins of the theory of curves lie in the nineteenth century work on complex function theory by Riemann, Abel, and Jacobi. Indeed, in some sense the theory of projective curves over the complex numbers is equivalent to the theory of compact Riemann surfaces, and one could learn a fair amount about Riemann surfaces by specializing results in this book, which are by and large valid over an arbitrary ground field k, to the case $k = \mathbb{C}$. To do so, however, would be a big mistake for two reasons. First, some of our results, which are obtained with considerable difficulty over a general field, are much more transparent and intuitive in the complex case. Second, the topological structure of complex curves and their beautiful relationship to complex function theory are very important parts of the subject that do not seem to generalize to arbitrary ground fields. The complex case in fact deserves a book all to itself, and indeed there are many such, e.g. [15].

The generalization to arbitrary gound fields is a twentieth century development, pioneered by the German school of Hasse, Schmidt, and Deuring in the 1920s and 1930s. A significant impetus for this work was provided by the development of algebraic number theory in the early part of the century, for it turns out that there is a very close analogy between algebraic function fields and algebraic number fields.

The results of the German school set the stage for the development of algebraic geometry over arbitrary fields, but were in large part limited to the special case of curves. Even in that case, there were serious difficulties. For example, Hasse was able to prove the Riemann hypothesis only for elliptic curves. The proof for curves of higher genus came from Weil and motivated his breakthrough work on abstract varieties. This in turn led to the "great leap forward" by the French school of Serre, Grothendiek, Deligne, and others to the theory of schemes in the 1950s and 1960s.

The flowering of algebraic geometry in the second half of the century has, to a large extent, subsumed the theory of algebraic curves. This development has been something of a two-edged sword, however. On the one hand, many of the results on curves can be seen as special cases of more general facts about schemes. This provides the usual benefits of a unified and in some cases a simplified treatment, together with some further insight into what is going on. In addition, there are some important facts about curves that, at least with the present state of knowledge, can only be understood with the more powerful tools of algebraic geometry. For example, there are important properties of the Jacobian of a curve that arise from its structure as an algebraic group.

On the other hand, the full-blown treatment requires the student to first master the considerable machinery of sheaves, schemes, and cohomology, with the result that the subject becomes less accessible to the nonspecialist. Indeed, the older algebraic development of Hasse et al. has seen something of a revival in recent years, due in part to the emergence of some applications in other fields of mathematics such as cryptology and coding theory. This approach, which is the one followed in this book, treats the function field of the curve as the basic object of study.

In fact, one can go a long way by restricting attention entirely to the function field (see, e.g., [17]), because the theory of function fields turns out to be equivalent to the theory of nonsingular projective curves. However, this is rather restrictive because many important examples of projective curves have singularities. A feature of this book is that we go beyond the nonsingular case and study projective curves in general, in effect viewing them as images of nonsingular curves.

What Is an Algebraic Function?

For our purposes, an algebraic function field K is a field that has transcendence degree one over some base field k, and is also finitely generated over k. Equivalently, K is a finite extension of $k(x)$ for some transcendental element $x \in K$. Examples of such fields abound. They can be constructed via elementary field theory by simply adjoining to $k(x)$ roots of irreducible polynomials with coefficients in $k(x)$. In addition, however, we will always assume that k is the full field of constants of K, that is, that every element of K that is algebraic over k is already in k.

When k is algebraically closed, there is another more geometric way to construct such fields, which is more closely related to the subject of this book. Let $\mathbb{P}^2$ be the set of lines through the origin in complex 3-space, and let $V \subseteq \mathbb{P}^2$ be a projective curve as described above. That is, V is the set of zeros of a complex, irreducible, homogenous polynomial $f(X_0,X_1,X_2)$ modulo scalar equivalence. We observe that a quotient of two homogeneous polynomials of the same degree defines a complex-valued function at all points of $\mathbb{P}^2$ where the denominator does not vanish. If the denominator does not vanish identically on V, it turns out that restricting this function to V defines a complex-valued function at all but a finite number of points of V. The set of all such functions defines a subfield $\mathbb{C}(V)$, which is called the *function field* of V.

Of course, there is nothing magical about the complex numbers in this discussion — any algebraically closed field k will do just as well. In fact, every finitely generated extension K of an algebraically closed field k of transcendence degree one arises in this way as the function field of a projective nonsingular curve V defined over k which, with suitable definitions, is unique up to isomorphism. This explains why we call such fields "function fields", at least in the case when k is algebraically closed.

What Is in This Book?

Here is a brief outline of the book, with only sketchy definitions and of course no proofs.

It turns out that for almost all points P of an algebraic curve V, the order of vanishing of a function at P defines a discrete k-valuation v_P on the function field K of V. The valuation ring $\mathcal{O}_P$ defined by v_P has a unique maximal ideal I_P, which, because v_P is discrete, is a principal ideal. A generator for I_P is called a *local parameter* at P. It is convenient to identify I_P with P. Indeed, for the first three chapters of the book, we forget all about the curve V and its points and focus attention instead on the set $\mathbb{P}_K$ of k-valuation ideals of K, which we call the set of *prime divisors* of K. A basic fact about function fields is that all k-valuations are discrete.

A *divisor* on the function field K is an element of the free abelian group $\mathrm{Div}(K)$ generated by the prime divisors. There is a map $\deg : \mathrm{Div}(K) \to \mathbb{Z}$ defined by $\deg(P) = |\mathcal{O}_P/P : k|$ for every prime divisor P. For $x \in K$, it is fundamental that the divisor

$$[x] = \sum_P v_P(x)P$$

has degree zero, and of course that the sum is finite. In other words, every function has the same (finite) number of poles and zeros, counting multiplicities. Divisors of the form $[x]$ for some $x \in K$ are called *principal divisors* and form a subgroup of $\mathrm{Div}(K)$.

A basic problem in the subject is to construct a function with a given set of poles and zeros. Towards this end, we denote by $\leq$ the obvious partial order on $\mathrm{Div}(K)$, and we define for any divisor D,

$$L(D) := \{x \in K \mid [x] \geq -D\}.$$

So for example if S is a set of distinct prime divisors and D is its sum, $L(D)$ is the set of all functions whose poles lie in the set S and are simple.

It is elementary that $L(D)$ is a k-subspace of dimension at most $\deg(D)+1$. The fundamental theorem of Riemann asserts the existence of an integer g_K such that for all divisors D of sufficiently large degree, we have

$$(*) \qquad \dim_k(L(D)) = \deg(D) - g_K + 1.$$

The integer g_K is the *genus* of K. In the complex case, this number has a topological interpretation as the number of holes in the corresponding Riemann surface. A refinement of Riemann's theorem due to Roch identifies the error term in $(*)$ for divisors of small degree and shows that the formula holds for all divisors of degree at least $2g-1$.

Our proof of the Riemann–Roch theorem is due to Weil [23], and involves the expansion of a function in a formal Laurent series at each prime divisor. In the complex case, these series have a positive radius of convergence and can be integrated. In the general case, there is no notion of convergence or integration. It is an amazing fact, nevertheless, that a satisfactory theory of differential forms exists in general. Although they are not functions, differential forms have poles and zeros and therefore divisors, which are called *canonical divisors*. Not only that, they have residues that sum to zero, just as in the complex case. Our treatment of the residue theorem follows Tate [20].

There are also higher derivatives, called Hasse derivatives, which present some technical difficulties in positive characteristic due to potential division by zero. This topic seems to have been somewhat neglected in the literature on function fields. Our approach is based on Hensel's lemma. Using the Hasse derivatives, we prove the analogue of Taylor's theorem for formal power series expansion of a function in powers of a local parameter. This material is essential later on when we study Weierstrass points of projective maps.

Thus far, the only assumption required on the ground field k is that it be the full field of constants of K. If k is perfect (e.g. of characteristic zero, finite, or algebraically closed), this assumption suffices for the remainder of the book. For imperfect ground fields, however, technical difficulties can arise at this point, and we must strengthen our assumptions to ensure that $k' \otimes_k K$ remains a field for every finite extension k'/k. Then the space Ω_K of differential forms on K has the structure of a (one-dimensional!) K-vector space, which means that all canonical divisors are congruent modulo principal divisors, and thus have the same degree (which turns out to be $2g-2$).

Given a finite, separable extension K' of K, there is a natural map

$$K' \otimes_K \Omega_K \to \Omega_{K'},$$

which is actually an isomorphism. This allows us to compare the divisor of a differential form on K with the divisor of its image in K', and leads to the Riemann–Hurwitz formula for the genus:

$$2g_{K'} - 2 = \frac{|K' : K|}{|k' : k|}(2g_K - 2) + \deg \mathscr{D}_{K'/K}.$$

Here, the divisor $\mathscr{D}_{K'/K}$ is the *different*, an important invariant of the extension, and k' is the relative algebraic closure of k in K'. The different, a familiar object in algebraic number fields, plays a similar key role in function fields. The formula has many applications, e.g., in the *hyperelliptic* case, where we have $K = k(x)$ and $|K' : K| = 2$.

At this point, further technical difficulties can arise for general ground fields of finite characteristic, and to ensure, for example, that $\mathscr{D}_{K'/K} \geq 0$, we must make the additional technical assumption that all prime divisors are *nonsingular*. Fortunately, it turns out that this condition is always satisfied in some finite (purely inseparable!) scalar extension of K.

When k is not algebraically closed, the question of whether K has any prime divisors of degree one (which we call *points*) is interesting. There is a beautiful answer for k finite of order q, first proved for genus one by Hasse and in general by Weil. Let $a_K(n)$ denote the number of nonnegative divisors of K of degree n, and put

$$Z_K(t) = \sum_{n=1}^{\infty} a_K(n)t^n.$$

Note that $a_K(1)$ is the number of points of K. Following Stör–Voloch [19] and Bombieri [2], we prove that

$$Z_K(t) = \frac{1}{(1-t)(1-qt)} \prod_{i-1}^{2g}(1 - \alpha_i t),$$

where $|\alpha_i| = \sqrt{q}$. This leads directly to the so-called "Weil bound" for the number of points of K:

$$|a_K(1) - q - 1| \leq 2g\sqrt{q}.$$

Turning our attention now to projective curves, we assume that the ground field k is algebraically closed, and we define a closed subset of projective space to be the set of all zeros of a (finite) set of homogeneous polynomials. A projective variety is an irreducible closed set (i.e., not the union of two proper closed subsets), and a projective curve is a projective variety whose field of rational functions has transcendence degree one.

Given a projective curve $V \subseteq \mathbb{P}^n$, we obtain its function field K by restricting rational functions on $\mathbb{P}^n$ to V. To recover V from K, let $X_0, \ldots, X_n$ be the coordinates of $\mathbb{P}^n$ with notation chosen so that X_0 does not vanish on V. Then the rational functions $\phi_i := X_i/X_0$, $(i = 1, \ldots, n)$ are defined on V. Given a point P of K, we let $e_P = -\min_i\{v_P(\phi_i)\}$ and put

$$\phi(P) := (t^{e_P}\phi_0(P) : t^{e_P}\phi_1(P) : \cdots : t^{e_P}\phi_n(P)) \in \mathbb{P}^n,$$

where t is a local parameter at P. It is not hard to see that the image of ϕ is V. In fact, any finite dimensional k-subspace $L \subseteq K$ defines a map ϕ_L to projective space in this way whose image is a projective curve.

The map ϕ is always surjective. But when is it injective? This question leads us to the notion of singularities. Let $\phi(P) = a \in \mathbb{P}^n$, and let $\mathcal{O}_a$ be the subring of K consisting of all fractions f/g where f and g are homogeneous polynomials of the same degree and $g(a) \neq 0$. We say that ϕ is *nonsingular* at P if $\mathcal{O}_a = \mathcal{O}_P$. This is equivalent to the familiar condition that the matrix of partial derivatives of the coordinate functions be of maximal rank.

An everywhere nonsingular projective map is called a *projective embedding*. It turns out that $\phi_{L(D)}$ is an embedding for any divisor D of degree at least $2g+1$. Another interesting case is the *canonical map* $\phi_{L(D)}$ where D is a canonical divisor. The canonical map is an embedding unless K is hyperelliptic.

The study of singularities is particularly relevant for plane curves. We prove that a nonsingular plane curve of degree d has genus $(d-1)(d-2)/2$, so there are many function fields for which every map to P^2 is singular, e.g. any function field of genus 2. In fact, for a plane curve of degree d and genus g, we obtain the formula

$$g = \frac{(d-1)(d-2)}{2} - \frac{1}{2}\sum_Q \delta(Q),$$

where for each singularity Q, $\delta(Q)$ is a positive integer determined by the local behavior of V at Q.

All of the facts discussed above, and many more besides, are proved in this book. We have tried hard to make the treatment as self-contained as possible. To this end, we have also included an appendix on elementary field theory.

Finally, there is a website for the book located at *http://www.functionfields.org*. There you will find the latest errata, a discussion forum, and perhaps answers to some selected exercises.

1

Background

This chapter contains some preliminary definitions and results needed in the sequel. Many of these results are quite elementary and well known, but in the self-contained spirit of the book, we have provided proofs rather than references. In this book the word "ring" means "commutative ring with identity," unless otherwise explicitly stated.

1.1 Valuations

Let K be a field. We say that an integral domain $\mathcal{O} \subseteq K$ is a *valuation ring* of K if $\mathcal{O} \neq K$ and for every $x \in K$, either x or x^{-1} lies in $\mathcal{O}$. In particular, K is the field of fractions of $\mathcal{O}$. Thus, we call an integral domain $\mathcal{O}$ a valuation ring if it is a valuation ring of its field of fractions.

Given a valuation ring $\mathcal{O}$ of K, let $V = K^\times/\mathcal{O}^\times$ where for any ring R, $R^\times$ denotes the group of units of R. The *valuation* afforded by $\mathcal{O}$ is the natural map $v : K^\times \to V$. Although it seems natural to write V multiplicatively, we will follow convention and write it additively. We call V the group of values of $\mathcal{O}$. By convention, we extend v to all of K by defining $v(0) = \infty$.

For elements $a\mathcal{O}^\times, b\mathcal{O}^\times$ of V, define $a\mathcal{O}^\times \leq b\mathcal{O}^\times$ if $a^{-1}b \in \mathcal{O}$, and put $v < \infty$ for all $v \in V$. Then it is easy to check that the relation $\leq$ is well defined, converts V to a totally ordered group, and that

$$v(a+b) \geq \min\{v(a), v(b)\} \tag{1.1.1}$$

for all $a, b \in K^\times$.

Let $P := \{x \in \mathcal{O} \mid v(x) > 0\}$. Then P is the set of nonunits of $\mathcal{O}$. From (1.1.1), it follows that P is an ideal, and hence the unique maximal ideal of $\mathcal{O}$. If $v(a) > v(b)$, then $ab^{-1} \in P$, whence $v(1+ab^{-1}) = 0$ and therefore $v(a+b) = v(b)$. To summarize:

Lemma 1.1.2. *If $\mathcal{O}$ is a valuation ring with valuation v, then $\mathcal{O}$ has a unique maximal ideal $P = \{x \in \mathcal{O} \mid v(x) > 0\}$ and* (1.1.1) *is an equality unless, perhaps, $v(a) = v(b)$.* □

Given a valuation ring $\mathcal{O}$ of a field K, the natural map $K^\times \to K^\times/\mathcal{O}^\times$ defines a valuation. Conversely, given a nontrivial homomorphism v from $K^\times$ into a totally ordered additive group G satisifying $v(a+b) \geq \min\{v(a), v(b)\}$, we put $\mathcal{O}_v := \{x \in K^\times \mid v(x) \geq 0\} \cup \{0\}$. Then it is easy to check that $\mathcal{O}_v$ is a valuation ring of K and that v induces an order-preserving isomorphism from $K^\times/\mathcal{O}^\times$ onto its image. Normally, we will identify these two groups. Note, however, that some care needs to be taken here. If, for example, we replace v by $nv : K^\times \to G$ for any positive integer n, we get the same valuation of K.

We let $P_v := \{x \in K \mid v(x) > 0\}$ be the maximal ideal of $\mathcal{O}_v$ and $F_v := \mathcal{O}_v/P_v$ be the *residue field* of v. If K contains a subfield k, we say that v is a k-valuation of K if $v(x) = 0$ for all $x \in k^\times$. In this case, F_v is an extension of k. Indeed, in the case of interest to us, this extension turns out to be finite. However, there is some subtlety here because the residue fields do not come equipped with any particular fixed embedding into some algebraic closure of k, except in the (important) special case $F_v = k$.

Our first main result on valuations is the extension theorem, but first we need a few preliminaries.

Lemma 1.1.3. *Let R be a subring of a ring S and let $x \in S$. Then the following conditions are equivalent:*

1. *x satisfies a monic polynomial with coefficients in R,*
2. *$R[x]$ is a finitely generated R-module,*
3. *x lies in a subring that is a finitely generated R-submodule.*

Proof. The implications (1) ⇒ (2) ⇒ (3) are clear. To prove (3) ⇒ (1), let $\{x_1, \ldots, x_n\}$ be a set of R-module generators for a subring S_0 containing x, then there are elements $a_{ij} \in R$ such that

$$xx_i = \sum_{j=1}^{n} a_{ij}x_j \quad \text{for } 1 \leq i \leq n.$$

Multiplying the matrix $(\delta_{ij}x - a_{ij})$ by its transposed matrix of cofactors, we obtain

$$f(x)x_j = 0 \quad \text{for all } j,$$

where $f(X)$ is the monic polynomial $\det(\delta_{ij}X - a_{ij})$ and δ_{ij} is the Kronecker symbol. We conclude that $f(x)S_0 = 0$, and since $1 \in S_0$, that $f(x) = 0$. □

Given rings $R \subseteq S$ and $x \in S$, we say that x is *integral* over R if any of the above conditions is satisfied. We say that S is integral over R if every element of S is integral over R. If $R[x]$ and $R[y]$ are finitely generated R-modules with generators $\{x_i\}$ and $\{y_j\}$ respectively, it is easy to see that $R[x,y]$ is generated by $\{x_i y_j\}$. Then using (1.1.3) it is straightforward that the sum and product of integral elements is again integral, so the set $\hat{R}$ of all elements of S integral over R is a subring. Furthermore, if $x \in S$ satisfies

$$x^n + \sum_{i=0}^{n-1} a_i x^i = 0$$

with $a_i \in \hat{R}$, then x is integral over $\hat{R}_0 := R[a_0, \ldots, a_{n-1}]$, which is a finitely generated R-module by induction on n. If $\{b_1, \ldots, b_m\}$ is a set of R-module generators for $\hat{R}_0$, then $\{b_i x^j \mid 1 \le i \le m,\ 0 \le j < n\}$ generates $\hat{R}_0[x]$ as an R-module, and we have proved

Corollary 1.1.4. *The set of all elements of S integral over R forms a subring $\hat{R}$, and any element of S integral over $\hat{R}$ is already in $\hat{R}$.* □

The ring $\hat{R}$ is called the *integral closure* of R in S. If $\hat{R} = R$, we say that R is *integrally closed* in S. If S is otherwise unspecified, we take it to be the field of fractions of R.

Recall that a ring R is called a *local ring* if it has an ideal M such that every element of $R \setminus M$ is a unit. Then M is evidently the unique maximal ideal of R, and conversely, a ring with a unique maximal ideal is local. If R is any integral domain with a prime ideal P, the *localization* R_P of R at P is the (local) subring of the field of fractions consisting of all x/y with $y \notin P$.

Lemma 1.1.5 (Nakayama's Lemma). *Let R be a local ring with maximal ideal P and let M be a nonzero finitely generated R-module. Then $PM \subsetneq M$.*

Proof. Let $M = Rm_1 + \cdots + Rm_n$, where n is minimal, and put $M_0 := Rm_2 + \cdots + Rm_n$. Then M_0 is a proper submodule. If $M = PM$, we can write

$$m_1 = \sum_{i=1}^{n} a_i m_i$$

with $a_i \in P$, but $1 - a_1$ is a unit since R is a local ring, and we obtain the contradiction

$$m_1 = (1 - a_1)^{-1} \sum_{i=2}^{n} a_i m_i \in M_0.$$

□

Theorem 1.1.6 (Valuation Extension Theorem). *Let R be a subring of a field K and let P be a nonzero prime ideal of R. Then there exists a valuation ring $\mathcal{O}$ of K with maximal ideal M such that $R \subseteq \mathcal{O} \subseteq K$ and $M \cap R = P$.*

Proof. Consider the set of pairs (R',P') where R' is a subring of K and P' is a prime ideal of R'. We say that (R'',P'') *extends* (R',P') and write $(R'',P'') \geq (R',P')$ if $R'' \supseteq R'$ and $P'' \cap R' = P'$. This relation is a partial order. By Zorn's lemma, there is a maximal extension $(\mathcal{O},M)$ of (R,P).

We first observe that $M \neq 0$, so $\mathcal{O} \neq K$. Furthermore, after verifying that $M = M\mathcal{O}_M \cap \mathcal{O}$ we have $(\mathcal{O}_M, M\mathcal{O}_M) \geq (\mathcal{O},M)$. By our maximal choice of $(\mathcal{O},M)$ we conclude that $\mathcal{O}$ is a local ring with maximal ideal M. Now let $x \in K$. If M generates a proper ideal M_1 of $\mathcal{O}[x^{-1}]$, then $(\mathcal{O}[x^{-1}],M_1) \geq (\mathcal{O},M)$ because M is a maximal ideal of $\mathcal{O}$, and the maximality of $(\mathcal{O},M)$ implies that $x^{-1} \in \mathcal{O}$. Otherwise, there exists an integer n and elements $a_i \in M$ such that

$$(*) \qquad 1 = \sum_{i=0}^{n} a_i x^{-i}.$$

Since $\mathcal{O}$ is a local ring, $1 - a_0$ is a unit. Dividing $(*)$ by $(1-a_0)x^{-n}$, we find that x is integral over $\mathcal{O}$. In particular, $\mathcal{O}[x]$ is a finitely generated $\mathcal{O}$-module. Now the maximality of $(\mathcal{O},M)$ and (1.1.5) imply that $x \in \mathcal{O}$. □

Corollary 1.1.7. *Suppose that $k \subseteq K$ are fields and $x \in K$. If x is transcendental over k, there exists a k-valuation ν of K with $\nu(x) > 0$. If x is algebraic over k, $\nu(x) = 0$ for all k-valuations ν.*

Proof. If x is transcendental over k, apply (1.1.6) with $\mathcal{O} := k[x]$ and $P := (x)$ to obtain a k-valuation ν with $\nu(x) > 0$. Conversely, if

$$\sum_{i=0}^{n} a_i x^i = 0$$

with $a_i \in k$ and $a_n \neq 0$, and if ν is a k-valuation, then we have

$$\nu(a_n x^n) = n\nu(x) = \nu\Big(\sum_{i<n} a_i x^i\Big).$$

If $\nu(x)$ were nonzero, the right-hand side would be a sum of terms each of different value, and we would have $n\nu(x) = i\nu(x)$ for some i by repeated application of (1.1.2), which is impossible. Hence, $\nu(x) = 0$ as required. □

Corollary 1.1.8. *Let R be a subring of a field K. Then the intersection of all valuation rings of K containing R is the integral closure of R in K.*

Proof. If $x \in K$ satisfies a monic polynomial of degree n over R and ν is a valuation of K that is nonnegative on R, then there are $r_i \in R$ such that

$$n\nu(x) = \nu(x^n) = \nu\left(\sum_{i=0}^{n-1} r_i x^i\right) \geq \min_{0 \leq i < n} i\nu(x),$$

from which it follows that $\nu(x) \geq 0$. This shows that the integral closure is contained in the intersection.

To obtain equality, suppose that $x \in R[x^{-1}]$. Then there are $r_i \in R$ such that

$$x = \sum_{i=0}^{n} r_i x^{-i},$$

and multiplying through by x^n we see that x is integral over R. If, therefore, x is not integral over R, there is a maximal ideal P of $R[x^{-1}]$ containing x^{-1} and then by (1.1.6) there is a valuation of K that is positive at x^{-1} and hence negative at x. □

Lemma 1.1.9. *Let $\mathcal{O}$ be a valuation ring. Then finitely generated torsion-free $\mathcal{O}$-modules are free. In particular, finitely generated ideals are principal.*

Proof. Let P be a torsion-free $\mathcal{O}$-module with generating set $\{m_1,\ldots,m_n\}$. Supposing there to be a relation $\sum_i a_i m_i = 0$ where not all a_i are zero, we may choose notation so that $\nu(a_n) = \min\{\nu(a_i) \mid a_i \neq 0\}$. Put $b_i := a_i/a_n \in \mathcal{O}$. Then $m_n = -\sum_{i<n} b_i m_i$, which implies that P is generated by $\{m_1,\ldots,m_{n-1}\}$. The result follows by an obvious induction argument. □

We now specialize to the case of a valuation whose group of values is infinite cyclic. Such a valuation ν is called a *discrete valuation* and its valuation ring $\mathcal{O}_\nu$ is called a *discrete valuation ring*. We usually identify the value group of a discrete valuation with the integers. Any element of $\mathcal{O}_\nu$ of value 1 is called a *local parameter* at ν (or sometimes a local parameter at P_ν). Equivalently, a local parameter is just a generator for P_ν.

Lemma 1.1.10. *Let t be an element of a subring $\mathcal{O}$ of a field K. Then $\mathcal{O}$ is a discrete valuation ring of K with local parameter t if and only if every element $x \in K$ can be written $x = ut^i$ for some unit $u \in \mathcal{O}$.*

Proof. If every element of K is of the form ut^i, put $\mathcal{O}_0 := \{ut^i \in K \mid i \geq 0\} \subseteq \mathcal{O}$. It is obvious that $\mathcal{O}_0$ is both a valuation ring of K and a maximal subring of K, and that $K^\times/\mathcal{O}_0^\times$ is infinite cyclic. We conclude that $\mathcal{O} = \mathcal{O}_0$ is a discrete valuation ring of K with local parameter t.

Conversely, if $\mathcal{O}$ is a discrete valuation ring of K with local parameter t affording the valuation ν, let $x \in K$ and let $i := \nu(x)$. Then $\nu(x^{-1}t^i) = 0$, so $x^{-1}t^i = u$ is a unit. □

The following corollary is immediate.

Corollary 1.1.11. *Let $\mathcal{O}$ be a discrete valuation ring of K. Then $\mathcal{O}$ is a maximal subring of K, and if t is a local parameter, every ideal of $\mathcal{O}$ is generated by a power of t.* □

The next result is a special case of the fundamental structure theorem for finitely generated modules over a principal ideal domain, but since this case is somewhat simpler than the general case, we outline a proof here.

Theorem 1.1.12 (Smith Normal Form). *Let $\mathcal{O}$ be a discrete valuation ring with local parameter t and let A be a matrix with entries in $\mathcal{O}$. Then there exist matrices U, V with entries in $\mathcal{O}$ and unit determinant, and nonnegative integers*

$$e_1 \leq e_2 \leq \cdots \leq e_r,$$

such that UAV has (i,i)-entry equal to t^{e_i} for $1 \leq i \leq r$ and all other entries zero.

Proof. If $A = 0$, there is nothing to prove. Otherwise, multiplying by permutation matrices as necessary, we may assume that $e_1 := v(a_{11}) \leq v(a_{ij})$ for all i, j. Multiplying row 1 by a unit, we may assume that $a_{11} = t^{e_1}$.

Next, using elementary row and column operations as necessary, we can assume that $a_{1j} = a_{i1} = 0$ for $i, j \geq 2$. Now apply induction to the submatrix of A obtained by deleting the first row and column, and the result follows. □

Corollary 1.1.13. *Let $\mathcal{O}$ be a discrete valuation ring with local parameter t, let M be a free $\mathcal{O}$-module of finite rank, and let $N \subseteq M$ be a nonzero submodule. Then N is free, and there exists a basis $\{x_1, \ldots, x_n\}$ for M, a positive integer $r \leq n$, and nonnegative integers $e_1 \leq e_2 \leq \cdots \leq e_r$ such that $\{t^{e_1}x_1, t^{e_2}x_2, \ldots, t^{e_r}x_r\}$ is a basis for N.*

Proof. We first argue by induction on the rank of M that N is finitely generated. If M has rank one, this follows from (1.1.11). If M has rank $n > 1$, let M_0 be a free submodule of rank $n-1$. Then $N \cap M_0$ and $N/(N \cap M_0)$ are finitely generated by induction, whence N is finitely generated.

Next, choose any basis for M, and any finite set of generators for N. Let A be the matrix whose columns are the generators for N expressed with respect to the chosen basis for M. Apply (1.1.12). The matrix U defines a new basis $\{x_1, \ldots, x_n\}$ for M, and the matrix V defines a new set of generators for N, namely $\{t^{e_1}x_1, t^{e_2}x_2, \ldots, t^{e_r}x_r\}$. It is evident that there are no nontrivial $\mathcal{O}$-linear relations among the $t^{e_i}x_i$, and thus they are a basis for N. □

Here is the standard example of a discrete valuation. Let R be a unique factorization domain, and let $p \in R$ be a prime element. For $x \in R$, write $x = p^e x_0$ where $p \nmid x_0$ and put $v_p(x) = e$. Extend v_p to the field of fractions by $v_p(x/y) = v_p(x) - v_p(y)$. It is immediate that $\mathcal{O}_{v_p}$ is just the local ring $R_{(p)}$. We call v_p the *p-adic* valuation of R. In particular, it turns out that for the field of rational functions in one variable, essentially all valuations are p-adic.

Theorem 1.1.14. *Let v be a valuation of $K := k(X)$. Then either $v = v_p$ for some irreducible polynomial $p \in k[X]$, or $v(f(X)/g(X)) = \deg(g) - \deg(f)$, where f and g are any polynomials.*

Proof. If $v(X) \geq 0$, then $k[X] \subseteq \mathcal{O}_v$ and $P_v \cap k[X]$ is a prime ideal (p) for some irreducible polynomial p. This implies that the localization $k[X]_{(p)}$ lies in $\mathcal{O}_v$. But by the above discussion, $k[X]_{(p)}$ is a discrete valuation ring of $k(X)$. By (1.1.10), $k[X]_{(p)}$ is a maximal subring of $k(X)$, so $v = v_p$. Note that $v_p(X^{-1}) = 0$ unless $(p) = (X)$. Thus, if $v(X) < 0$, we replace X by X^{-1}, repeat the above argument, and conclude that $\mathscr{O}_\nu = k[X^{-1}]_{(X^{-1})}$. In particular, $\nu(X) = -1$, whence $\nu(f) = -\deg(f)$ for any polynomial $f \in k[X]$ by (1.1.2). □

Given such a nice result for $k(X)$, we might wonder what can be said about $k(X,Y)$. Unfortunately, once we enter the world of higher dimensions, the landscape turns very bleak indeed. See Exercise (1.1).

We now turn to our second main result on valuations, the weak approximation theorem. In order to understand this terminology, several remarks are in order. Given a discrete valuation ν on a field K, choose any convenient real number $b > 1$ and define $|x|_\nu := b^{-\nu(x)}$ for all $x \in K$. Then it is straightforward to verify that $|x|_\nu$ defines a norm on K, with the strong triangle inequality:

$$|x+y|_\nu \leq \max(|x|_\nu, |y|_\nu).$$

Hence the statement $\nu(x-y) \gg 0$ may be thought of as saying that x and y are very close to each other. We will pursue this idea more fully in the next section.

Lemma 1.1.15. *Let $\{\nu_1,\ldots,\nu_n\}$ be a set of distinct discrete valuations of a field K, and let m be a positive integer. Then there exists $e \in K$ such that $\nu_1(e-1) > m$ and $\nu_i(e) > m$ for $i > 1$.*

Proof. We first find an element $x \in K$ such that $\nu_1(x) > 0$ and $\nu_i(x) < 0$ for $i > 1$. Namely, if $n = 2$, we choose $x_i \in \mathscr{O}_{\nu_i} \setminus \mathscr{O}_{\nu_{3-i}}$ for $i = 1,2$. This is possible since $\mathscr{O}_{\nu_i}$ is a maximal subring of K by (1.1.10). Then $x := x_1/x_2$ has the required properties. For $n > 2$, we may assume by induction that x' has been chosen with $\nu_1(x') > 0$ and $\nu_i(x') < 0$ for $1 < i < n$. If $\nu_n(x') < 0$, we put $x := x'$. Otherwise, choose y with $\nu_1(y) > 0$ and $\nu_n(y) < 0$, then we can find a suitably large positive integer r such that $\nu_i(y^r) \neq \nu_i(x')$ for any i. Now (1.1.2) implies that $x := x' + y^r$ has the required properties.

Finally, we observe that $\nu_1(x^m) \geq m$, $\nu_1(1+x^m) = 0$, and $\nu_i(1+x^m) = \nu_i(x^m) \leq -m$. It follows that the conclusions of the lemma are satisfied with

$$e := \frac{1}{1+x^{m+1}}. \quad \square$$

Theorem 1.1.16 (Weak Approximation Theorem). *Suppose that $\nu_1,\ldots,\nu_n$ are distinct discrete valuations of a field K, $m_1,\ldots,m_n$ are integers, and $x_1,\ldots,x_n \in K$. Then there exists $x \in K$ such that $\nu_i(x-x_i) = m_i$ for $1 \leq i \leq n$.*

Proof. Choose elements $a_i \in K$ such that $\nu_i(a_i) = m_i$ for all i, and let $m_0 := \max_i m_i$. Now choose an integer M such that

$$M + \min_{i,j}\{\nu_i(x_j), \nu_i(a_j)\} \geq m_0.$$

By (1.1.15) there are elements $e_i \in K$ such that $\nu_i(e_j - \delta_{ij}) > M$ for $1 \leq i,j \leq n$, where δ_{ij} is the Kronecker delta. Put $y := \sum_j e_j x_j$. Then for all i we have

$$\nu_i(y-x_i) = \nu_i\left(\sum_j (e_j - \delta_{ij})x_j\right) > M + \min_j \nu_i(x_j) \geq m_0.$$

Put $z := \sum_i e_i a_i$. Then as above we have $v_i(z - a_i) > m_0$, and hence $v_i(z) = v_i(z - a_i + a_i) = m_i$ for all i. The result now follows with $x := y + z$. □

Our first application of (1.1.16) is to determine the structure of the intersection of a finite number of discrete valuation rings of K. So for any finite set $\mathscr{V}$ of discrete valuations of a field K, and any function $m : \mathscr{V} \to \mathbb{Z}$, define

$$K(\mathscr{V}; m) = \{x \in K \mid v(x) \geq m(v) \text{ for all } v \in \mathscr{V}\}.$$

Corollary 1.1.17. *Suppose that K is a field, $\mathscr{V}$ is a finite set of discrete valuations of K, and that every valuation ring of K containing $\mathscr{O}_{\mathscr{V}} := K(\mathscr{V}; 0)$ is discrete. Then $\mathscr{O}_{\mathscr{V}}$ is a principal ideal domain and $I \subseteq \mathscr{O}_{\mathscr{V}}$ is a nonzero ideal if and only if $I = K(\mathscr{V}; m)$ for some nonnegative function m uniquely determined by I. Moreover, $\mathscr{O}_{\mathscr{V}}/K(\mathscr{V}; m)$ has an $\mathscr{O}_{\mathscr{V}}$-composition series consisting of exactly $m(v)$ composition factors isomorphic to F_v (as $\mathscr{O}_{\mathscr{V}}$-modules) for each $v \in \mathscr{V}$.*

Proof. From the definitions it is obvious that $\mathscr{O}_{\mathscr{V}}$ is a ring, that $K(\mathscr{V}; m)$ is an $\mathscr{O}_{\mathscr{V}}$-module for all m, and that $K(\mathscr{V}; m) \subseteq K(\mathscr{V}; m')$ for $m - m'$ nonnegative. In particular, $K(\mathscr{V}; m)$ is an ideal of $\mathscr{O}_{\mathscr{V}}$ for m nonnegative.

Conversely, let $0 \neq I \subseteq \mathscr{O}_{\mathscr{V}}$ be an ideal, and for each $v \in \mathscr{V}$ put

$$m(v) := \min_{x \in I} v(x).$$

By (1.1.16) there exists $x_m \in K$ with $v(x_m) = m(v)$ for all $v \in \mathscr{V}$. Then $x_m \in \mathscr{O}_{\mathscr{V}}$, and $x_m^{-1} I$ is an ideal of $\mathscr{O}_{\mathscr{V}}$ that is not contained in P_v for any $v \in \mathscr{V}$. If, by way of contradiction, $x_m^{-1} I \subsetneq \mathscr{O}_{\mathscr{V}}$, then (1.1.6) yields a valuation ring $\mathscr{O}_{v'}$ containing $\mathscr{O}_{\mathscr{V}}$ with $x_m^{-1} I \subseteq P_{v'}$. Thus, $v' \notin \mathscr{V}$, but by hypothesis v' is discrete. Now (1.1.16) yields an element $y \in \mathscr{O}_{\mathscr{V}}$ with $v'(y) < 0$, a contradiction. We conclude that $x_m^{-1} I = \mathscr{O}_{\mathscr{V}}$, i.e., $I = \mathscr{O}_{\mathscr{V}} x_m$ is principal. If $K(\mathscr{V}; m) = K(\mathscr{V}; m')$, then from $x_m \in K(\mathscr{V}; m')$ and $x_{m'} \in K(\mathscr{V}; m)$ we obtain

$$m(v) = v(x_m) \geq m'(v) = v(x_{m'}) \geq m(v)$$

for all $v \in \mathscr{V}$, whence $m = m'$.

In particular, the $\mathscr{O}_{\mathscr{V}}$-module $K(\mathscr{V}; m)/K(\mathscr{V}; m + \delta_v)$ is irreducible, where for $v \in \mathscr{V}$ we define

$$\delta_v(v') := \begin{cases} 1 & \text{for } v = v', \\ 0 & \text{otherwise.} \end{cases}$$

Let t be a local parameter at v. Then the map

$$\eta(x) := t^{-m(v)} x + P_v$$

defines an additive map $\eta : K(\mathscr{V}; m) \to F_v$ with $\ker \eta = K(\mathscr{V}; m + \delta_v)$. This map gives F_v an $\mathscr{O}_{\mathscr{V}}$ action, because as we next argue, η is surjective.

Namely, for $y \in \mathscr{O}_v$ (1.1.16) yields an element $x \in K$ with $v'(x) \geq m(v')$ for $v' \in \mathscr{V}$, $v' \neq v$ and $v(x - t^{m(v)} y) \geq m(v) + 1$. This implies that $x \in K(\mathscr{V}; m)$ and $\eta(x) \equiv y \mod P_v$, so η is surjective and induces an $\mathscr{O}_{\mathscr{V}}$-module isomorphism

$K(\mathscr{V};m)/K(\mathscr{V};m+\delta_v) \simeq F_v$. Now an obvious induction argument shows that $\mathscr{O}_{\mathscr{V}}/K(\mathscr{V};m)$ has a composition series consisting of exactly $m(v)$ composition factors isomorphic to F_v for each $v \in \mathscr{V}$. □

Corollary 1.1.18. *With the above notation, we have*

$$K(\mathscr{V};m)+K(\mathscr{V}';m') = K(\mathscr{V}\cap\mathscr{V}';\min\{m,m'\})$$

for m and m' nonnegative.

Proof. It is obvious that $K(\mathscr{V};m)+K(\mathscr{V}';m') \subseteq K(\mathscr{V}\cap\mathscr{V}';\min\{m,m'\})$. Conversely, let $y \in K(\mathscr{V}\cap\mathscr{V}';\min\{m,m'\})$. Write $y = ye + y(1-e)$, where e is chosen using (1.1.16) such that

$$\begin{aligned} v(e) &\geq m(v) - v(y) \text{ for } v \in \mathscr{V}\setminus\mathscr{V}', \\ v(e) &\geq m(v) \text{ for } v \in \mathscr{V}\cap\mathscr{V}' \text{ and } m(v) \geq m'(v), \\ v(1-e) &\geq m'(v) \text{ for } v \in \mathscr{V}\cap\mathscr{V}' \text{ and } m(v) < m'(v), \\ v(1-e) &\geq m'(v) - v(y) \text{ for } v \in \mathscr{V}'\setminus\mathscr{V}. \end{aligned}$$

We claim that $ye \in K(\mathscr{V};m)$, i.e. that $v(y)+v(e) \geq m(v)$ for all $v \in \mathscr{V}$. This is clear for $v \notin \mathscr{V}'$ and for $v \in \mathscr{V}\cap\mathscr{V}'$ with $m(v) \geq m'(v)$, because $v(y) \geq 0$ in this case. For $v \in \mathscr{V}\cap\mathscr{V}'$ with $m(v) < m'(v)$ we have $v(y) \geq m(v)$ and $v(1-e) \geq m'(v) \geq 0$, so $v(e) \geq 0$ as well, and thus all conditions are satisfied. Similarly, it follows that $y(1-e) \in K(\mathscr{V}';m')$. □

Our final results on valuations concern the behavior of a discrete valuation under a finite degree field extension. Suppose that v is a discrete valuation of K and K' is a finite extension of K. Then (1.1.6) shows that there exists a valuation ring $\mathscr{O}'$ of K' containing $\mathscr{O}_v$ whose maximal ideal contains P_v. If v' is the associated valuation of K', we say that v' *divides* v and write $v'|v$. We are tempted to write $v'|_K = v$, but some care must be taken with this statement, particularly since it turns out that v' is also discrete, and we are in the habit of identifying the value group of a discrete valuation with $\mathbb{Z}$. If we do this for both v and v', then what in fact happens is that $v'|_K = ev$ for some positive integer e.

Theorem 1.1.19. *Suppose that v is a discrete valuation of a field K, K' is a finite extension of K, and v' is a valuation of K' dividing v. Then v' is discrete, and there is a positive integer $e \leq |K':K|$ such that $v'|_K = ev$.*

Proof. Let $n = |K':K|$ and let V (resp. V') be the canonical group of values of v (resp. v'). That is, $V = K^\times/\mathscr{O}_v^\times$, and V' is defined similarly. For the remainder of this argument we will not identify either group with $\mathbb{Z}$. Then since $\mathscr{O}_{v'}^\times \cap K^\times = \mathscr{O}_v^\times$, we see that V is canonically isomorphic to a subroup of V', and $v'|_K = v$.

We argue that V has index at most n in V', for if not, there are values $\{v'_0, v'_1, \ldots, v'_n\} \subseteq V'$, no two of which differ by an element of V. Choose elements $x'_i \in K'$ such that $v'(x'_i) = v'_i$ for $0 \leq i \leq n$, then there is a dependence relation

$$\sum_{i=0}^{n} a_i x'_i = 0$$

with $a_i \in K$. Carefully clearing denominators, we may assume that the a_i are in $\mathcal{O}_\nu$ and at least one, say a_0, is nonzero. Note that by our choice of ν_i', we have

$$\nu'(a_i x_i') - \nu'(a_j x_j') - \nu(a_i) + \nu_i' - \nu(a_j) - \nu_j' \neq 0$$

for all $i \neq j$ for which a_i and a_j are nonzero. But now (1.1.2) implies that

$$\nu'(a_0 x_0') = \nu'(\sum_{i>0} a_i x_i') = \nu'(a_j x_j')$$

for that index $j > 0$ for which $\nu'(a_j x_j')$ is minimal. This contradiction shows that $|V' : V| \leq n$.

Let $e := |V' : V|$ and let a be a positive generator for V. There are at most e elements of V' in the interval $[0, a]$ since no two of them can be congruent modulo V. In particular, V' has a smallest positive element; call it b. Let $v' \in V'$. Then $ev' \in V$, and we get $v' \leq ev' \leq m'eb$ for some positive integer m'. Let m be the least positive integer for which $mb \geq v'$. Then $v' > (m-1)b$, and hence $0 \geq mb - v' > b$. By our choice of b we conclude that $v' = mb$ and thus that V' is cyclic as required. □

We call the integer $e = e(\nu'|\nu)$ of (1.1.19) the *ramification index* of ν' over ν. We will often write $e(P'|P)$ for $e(\nu'|\nu)$, where P (resp. P') is the valuation ideal of ν (resp. ν'). When $e > 1$ we say that P is *ramified* in K'.

Lemma 1.1.20. *Let $\mathcal{O}$ be a discrete valuation ring with field of fractions K, maximal ideal P, and residue field F. Let M be a torsion-free $\mathcal{O}$-module with $\dim_K K \otimes_{\mathcal{O}} M = n$. Then $\dim_F M/PM \leq n$ with equality if and only if M is finitely generated.*

Proof. If M is finitely generated, it is free by (1.1.9) and therefore free of rank n, whence $\dim_F M/PM = n$ as well.

Suppose that $x_1, x_2, \ldots, x_m \in M$. If we have a nontrivial dependence relation

$$\sum_{i=1}^{m} a_i x_i = 0$$

with $a_i \in K$, we can carefully clear denominators, obtaining a relation with $a_i \in \mathcal{O}$ but not all $a_i \in P$. It follows that if the x_i are linearly independent modulo PM, they are linearly independent over K, and therefore $\dim_F M/PM \leq n$.

Assume now that $\dim_F M/PM = n$. Then lifting a basis of M/PM to M, we obtain by the previous paragraph a linearly independent set of cardinality n, which therefore generates a free submodule $M_0 \subseteq M$ of rank n, with $M_0 + PM = M$. Let $m \in M$ and put $N := M_0 + \mathcal{O}m$. Then N is torsion-free and thus also free (see (1.1.9)). Since it contains a free submodule of rank n, and any free submodule of M can have rank at most n, N also has rank n. Now (1.1.13) yields a basis $x_1, \ldots, x_n$ for N and nonnegative integers $i_1 \leq i_2 \leq \cdots \leq i_n$ such that $t^{i_1}x_1, \ldots, t^{i_n}x_n$ is a basis for M_0, where t is a local parameter for P. However, since $(M_0 + PM)/PM \simeq M_0/(M_0 \cap PM)$ has rank n, all the i_j must be zero, and hence $N = M$. Since m was arbitrary, we have $M_0 = M$ as required. □

Lemma 1.1.21. *Let $|K' : K| = n$, let $\mathcal{O}_\nu$ be a discrete valuation ring of K, and let R be any subring of K' containing the integral closure of $\mathcal{O}_\nu$ in K'. Then the map*

$$K \otimes_{\mathcal{O}_\nu} R \to K'$$

sending $x \otimes y$ to xy is an isomorphism of K-vector spaces. In particular, if $\nu'|\nu$, then the residue field $F_{\nu'}$ is an extension of the residue field F_ν of ν of degree at most n.

Proof. We first argue that the map $x \otimes y \mapsto xy$ is an embedding. Let t be a local parameter for ν. Then any element of the kernel can be written $x = \sum_{i=0}^{n} t^{-e_i} \otimes x_i$, where notation can be chosen so that $e_0 = \max_i e_i$. Then $\sum_i t^{-e_i} x_i = 0$, and we have

$$t^{e_0} x = \sum_i t^{e_0 - e_i} \otimes x_i = 1 \otimes \left(\sum_i t^{e_0 - e_i} x_i \right) = 0,$$

and therefore $x = 0$. To show that the map is surjective, let $y \in K'$. Then

$$\sum_{i=0}^{n} a_i y^i = 0$$

for $a_i \in K$. Since K is the field of fractions of $\mathcal{O}_\nu$, we can clear denominators and assume $a_i \in \mathcal{O}_\nu$. Multiplying through by a_n^{n-1} we see that $a_n y$ is integral over $\mathcal{O}_\nu$ and therefore $z := a_n y \in R$. Since $y = z/a_n$ we have $K' = KR$ as required.

In particular, we have $\dim_K K \otimes_{\mathcal{O}_\nu} \mathcal{O}_{\nu'} = n$, and we obtain from (1.1.20) the inequalities

$$\dim_F(\mathcal{O}_{\nu'}/P_{\nu'}) \le \dim_F(\mathcal{O}_{\nu'}/P\mathcal{O}_{\nu'}) \le n. \quad \square$$

The degree of the residue field extension is called the *residue degree* of ν' over ν, denoted $f(\nu'|\nu)$, or sometimes $f(P'|P)$. We can now prove a basic result on finite extensions.

Theorem 1.1.22. *Let K' be a finite extension of K and let $\mathcal{O}$ be a discrete valuation ring of K with maximal ideal P and residue field F. Let $\{\mathcal{O}_1, \ldots, \mathcal{O}_r\}$ be distinct valuation rings of K' containing $\mathcal{O}$, and let R be their intersection. Let P_i be the maximal ideal of $\mathcal{O}_i$ and put $e_i := e(P_i|P)$ and $f_i := f(P_i|P)$ for each i. Then*

1. *R contains a local parameter t_i for $\mathcal{O}_i$, $P_i \cap R = t_i R$, and $\mathcal{O}_i = R + P_i$ for each $i = 1, \ldots, r$.*

2. *$\dim_F R/PR = \sum_{i=1}^{r} e_i f_i \le |K' : K|$ with equality if and only if R is a finitely generated $\mathcal{O}$-module.*

In particular, there are only finitely many distinct valuation rings of K' containing $\mathcal{O}$.

Proof. Let ν_i be the valuation afforded by $\mathcal{O}_i$ for all i, and let $\mathcal{V} = \{\nu_1, \ldots, \nu_r\}$. Note that $R = K(\mathcal{V}; 0)$ and that any valuation ring of K' containing R also contains

$\mathcal{O}$ and is therefore discrete by (1.1.19). Thus, (1.1.17) applies. By (1.1.17) R is a PID and the maximal ideals of R are just the ideals $K(\mathcal{V};\delta_i) = P_i \cap R$ for $1 \leq i \leq r$, where $\delta_i(v_j) := \delta_{ij}$. Moreover, each such maximal ideal is generated by an element t_i that must be a local parameter at P_i. Let $x \in \mathcal{O}_i$. Then there is an element $x' \in K'$ with $v_i(x - x') \geq 1$. Moreover, $v_j(x') \geq 0$ for $j \neq i$ by (1.1.16); hence $\mathcal{O}_i = R + P_i$, proving (1).

Let t be a local parameter for P. Then $v_i(t) = e_i$ for all i. Since $PR = Rt = K(\mathcal{V};e)$ for some uniquely determined function $e : \mathcal{V} \to \mathbb{Z}$ by (1.1.17), we must have $e(v_i) := e_i$ for all i. Now we see that R/PR has exactly e_i composition factors isomorphic to F_i, whence

$$\dim_F(R/PR) = \sum_{i=1}^{r} e_i f_i.$$

Since $\dim_K K \otimes_{\mathcal{O}} R = n$ by (1.1.21), (2) follows from (1.1.20). □

If we take the complete set of extensions of $\mathcal{O}$ to K' above, the ring R is the integral closure of $\mathcal{O}$ in K' by (1.1.8). We see that the question of whether or not $\sum_i e_i f_i = n$ is equivalent to another important issue: When is the integral closure of $\mathcal{O}$ in K' a finitely generated $\mathcal{O}$-module? In (2.1.17) we will show that for function fields of curves, both conditions are always satisified.

The next result gives a very useful sufficient condition for all extensions of a discrete valuation v to be unramified.

Theorem 1.1.23. *Suppose that $\mathcal{O}$ is a discrete valuation ring of K with maximal ideal P and residue field F. Let $\{(\mathcal{O}_i, P_i) \mid 1 \leq i \leq r\}$ be the set of distinct extensions of $(\mathcal{O}, P)$ to some finite extension K' of K of degree n, and let $F_i := \mathcal{O}_i/P_i$. If we can write $K' = K(y)$ for some element $y \in K'$ whose monic minimum polynomial $g(X)$ has coefficients in $\mathcal{O}$ and has distinct roots mod P, then $e(P_i|P) = 1$ for all i. Moreover, $\mathcal{O}[y]$ is the integral closure of $\mathcal{O}$ in K', and $g(X)$ factors over F as a product of r distinct irreducibles*

$$g(X) \equiv \prod_{i=1}^{r} g_i(X) \mod P,$$

where notation can be chosen so that $F_i \simeq F[X]/(g_i(X))$. In particular, $\deg g_i = f(P_i|P)$.

Proof. Since $g(X)$ has distinct roots mod P, there is certainly a factorization

$$\bar{g}(X) = \prod_{i=1}^{r'} g_i(X)$$

into distinct irreducibles over $F[X]$, where $\bar{g}(X)$ is the reduction of $g(X)$ mod P and $g_i \in F[X]$. The map $X \mapsto y$ defines an epimorphism $\phi : \mathcal{O}[X] \to \mathcal{O}[y]$ whose kernel contains the principal ideal $(g(X))$. Since $g(X)$ is monic, $\mathcal{O}[X]/(g(X))$ is free on the basis $\{1, X, \ldots, X^{n-1}\}$. On the other hand, ϕ is the restriction of a map $K[X] \to K'$ whose kernel is generated by $g(X)$, and therefore $\ker(\phi)/(g(X))$ is a torsion $\mathcal{O}$-module. This implies that $\ker(\phi) = (g(X))$, and thus,

$$(*) \qquad \mathcal{O}[y]/P\mathcal{O}[y] = F[\bar{y}] \simeq F[X]/(\bar{g}(X)) \simeq \bigoplus_{i=1}^{r'} F[X]/(g_i(X)),$$

where $\bar{y} := (y+P)/P$.

To get the last isomorphism, note that the polynomials $\{\bar{g}/g_i \mid 1 \le i \le r'\}$ are relatively prime, so there exist polynomials $h_i \in F[X]$ with

$$\sum_{i=1}^{r'} h_i \frac{\bar{g}}{g_i} = 1.$$

Put $e_i := h_i g/g_i$. Then it is easy to check that $e_i e_j \equiv \delta_{ij} \mod g(X)$, so the e_i give the required decomposition of $F[X]/(\bar{g}(X))$.

Since y is integral over $\mathcal{O}$, $\mathcal{O}[y] \subseteq R := \cap_i \mathcal{O}_i$ by (1.1.8). Now using (1.1.22) and $(*)$ we have

$$(**) \qquad n = \deg(g) = \sum_{i=1}^{r'} f_i' = \dim_F \mathcal{O}[y]/P\mathcal{O}[y] \le \dim_F R/PR = \sum_{i=1}^{r} e_i f_i \le n,$$

where $e_i := e(P_i|P)$, and $f_i := f(P_i|P)$. We conclude that all of the above inequalities are equalities. In particular, R is a finitely generated $\mathcal{O}$-module by (1.1.20), and $R = \mathcal{O}[y] + PR$. Now Nakayama's lemma (1.1.5) applied to the finitely generated $\mathcal{O}$-module $R/\mathcal{O}[y]$ yields $R = \mathcal{O}[y]$.

Moreover, we have

$$R/PR = \mathcal{O}[y]/P\mathcal{O}[y] \simeq \bigoplus_{i=1}^{r'} F[X]/g_i(X),$$

and it follows that R/PR has exactly r' distinct maximal ideals. Thus, (1.1.19) yields $r' = r$, and after a suitable renumbering, that $F_i \simeq F[X]/g_i(X)$ for each i. In particular, $f_i = f_i'$, and $(**)$ implies that $e_i = 1$ for each i. □

Whenever K' is a separable extension of K, we can find a primitive element y for which $K' = K(y)$ by (A.0.17). If y is not integral over $\mathcal{O}_P$, (1.1.21) shows that we can replace it by a K-multiple that is integral. Then the monic minimum polynomial of y will have coefficients in $\mathcal{O}_P$. The problem is that it may not have distinct roots $\mod P$. As we will see later, however, in the case of interest there are only finitely many P for which this happens. Thus, (1.1.23) can be thought of as the generic case.

In the opposite direction, we say that a discrete valuation ν of K is *totally ramified* in K' if $e(\nu'|\nu) = |K' : K|$ for some ν', which is then unique by (1.1.22).

Theorem 1.1.24. *Suppose that $|K' : K| = n$ and that ν is a discrete valuation of K with $e(\nu'|\nu) = n$ for some discrete valuation ν' of K'. Let s be a local parameter at ν. Then $K' = K(s)$, s is integral over $\mathcal{O}_\nu$, and $\mathcal{O}_{\nu'} = \mathcal{O}_\nu[s]$.*

Proof. For any n-tuple $\{a_0,\dots,a_{n-1}\}$ of elements of K, let I be the set of indices i for which $a_i \neq 0$. Then for all $i \in I$ we have $\nu'(a_i) \equiv 0 \mod n$ and thus $\nu'(a_i s^i) \equiv i \mod n$. In particular, the integers $\{\nu'(a_i s^i) \mid i \in I\}$ are distinct, whence

$$(*) \qquad \nu'\left(\sum_{i=0}^{n-1} a_i s^i\right) = \min_{i \in I} \nu'(a_i s^i),$$

provided that $I \neq \emptyset$. It follows that $S := \{1, s, \dots, s^{n-1}\}$ is linearly independent over K, and is therefore a K-basis for K'. Let $x \in \mathcal{O}_{\nu'}$ and write $x = \sum_i a_i s^i$ with $a_i \in K$. Then $(*)$ implies that $\nu'(a_i) \geq 0$ for all i. We conclude that $\mathcal{O}_{\nu'} = \mathcal{O}_\nu[s]$, and that s is integral over $\mathcal{O}_\nu$. □

We finally observe that the ramification index and residue degree are both multiplicative:

Lemma 1.1.25. *Suppose $K_0 \subseteq K_1 \subseteq K_2$ are three fields with $|K_2 : K_0| < \infty$, and ν_i is a discrete valuation of K_i $(0 \leq i \leq 2)$ with $\nu_2|\nu_1|\nu_0$. Then*

$$e(\nu_2|\nu_0) = e(\nu_2|\nu_1)e(\nu_1|\nu_0), \quad \textit{and}$$
$$f(\nu_2|\nu_0) = f(\nu_2|\nu_1)f(\nu_1|\nu_0).$$

Proof. The first statement is immediate from the definition of e and the fact that restriction of functions is transitive. The second statement follows from the natural inclusion of residue fields $F_0 \subseteq F_1 \subseteq F_2$ and (A.0.2). □

At this point, an example may be in order. Let $K := \mathbb{Q}(x)$ be the field of rational functions in x over the rational numbers $\mathbb{Q}$, and let $K' := K(y)$, where $y^2 = p(x) := x^3 + x - 1$. Note that $p(x)$ is irreducible over $\mathbb{Q}$ because it does not have a rational root. Moreover, $|K' : K| = 2$, and every element of K' can be uniquely written $a(x) + b(x)y$ where a, b are rational functions of x. For $u = a + by$, define $\bar{u} := a - by$, and $N(u) := u\bar{u} = a^2 - b^2y^2 \in K$. Then

$$(*) \qquad u^{-1} = \frac{1}{a^2 - b^2y^2}(a - by).$$

In the following discussion, ν will denote a valuation of K with valuation ring $\mathcal{O}$, maximal ideal P, and residue field F, while ν' will be an extension of ν to K' with corresponding notation $\mathcal{O}', P', F'$. We will look at the three cases $\nu = \nu_x$, ν_{x-1}, and ν_∞.

$\nu := \nu_x$:

Then $\mathcal{O} := \mathbb{Q}[x]_{(x)}$ is the ring of local integers at x, *i.e.* the rational functions with no pole at $x = 0$. In this case, we claim that ν' is unique, and is given by

$$\nu'(a + by) := \min\{\nu(a), \nu(b)\}.$$

To see this, first note that $2\nu'(y) = \nu'(y^2) = \nu'(x^3 + x - 1) \geq 0$, so $y \in \mathcal{O}'$. Moreover, $y^2 \equiv -1 \mod P'$, which shows that F' contains the field $Q(i)$.

By (1.1.22) we conclude that $f=2, e=1$, and v' is unique. Moreover, every element of K' can be uniquely written in the form $v := x^i(a(x)+b(x)y)$, where $a(x)$ and $b(x)$ are local integers, at least one of which is not divisible by x. Put $u := a(x)+b(x)y$. Then $u \in \mathscr{O}'$ and $N(u)(0) = a(0)^2+b(0)^2$ is nonzero at $x=0$, whence $(*)$ shows that u is a unit in $\mathscr{O}'$.

$v := v_\infty$:

Changing variables by replacing x by x_1^{-1}, y by y_1^{-1}, and v_∞ by $v_{(x_1)}$ and then dropping the subscripts, we look instead at the equation

$$y^2 = \frac{x^3}{1+x^2-x^3}$$

at $x=0$. We first notice that $2v'(y) = 3v'(x)$ because $v'(1+x^2-x^3) = 0$. This implies that $2v'(yx^{-1}) = v'(x) > 0$, from which it follows that yx^{-1} is a local integer and that $e>1$. By (1.1.22), we conclude that $e=2, f=1$, and v' is unique. Since $e=2$ and $2v'(yx^{-1}) = v'(x)$, we see that yx^{-1} is a local parameter at P'. So in order to describe $\mathscr{O}'$, we need to know how to write every element of K' as the product of a local unit and a power of yx^{-1}. In contrast to the previous case, this is not entirely obvious, and we will defer the discussion for the moment.

$v = v_{x-1}$:

This time, we have $y^2 \equiv 1 \mod P'$, so $y \equiv \pm 1$. There are two extensions of v here, and the choice of sign will distinguish them. More precisely, we have $y^2-1 = x^3+x-2 = (x-1)(x^2+x+2)$. This means that the subring $k[x,y] \subseteq K$ has proper ideals $(x-1, y-1)$ and $(x-1, y+1)$ [1], so the valuation extension theorem produces a valuation v' with $v'(y-1)>0$ and an algebraically conjugate valuation v'' with $v''(y+1)>0$. Now (1.1.22) says that $e=f=1$, so $v'(x-1)=1$ and $x-1$ is a local parameter. Again, as in the previous case, it is not obvious how to write every element of K' as the product of a local unit and a power of $x-1$.

In the last two cases above, the question remains of how to actually compute the valuation v', or at least how to tell whether an element $a(x)+b(x)y$ is a local integer. We will discuss the case $v = v_{x-1}$, since the other case is essentially similar. Of course, if $a(x)$ and $b(x)$ are both local integers, so is u. The problem is that a and b can have poles that are canceled by the zero of y, or just by subtraction. For example, the element

$$u = \frac{y-1}{x-1} = \frac{x^2+x+2}{y+1}$$

is a local integer with the value 2 at $(1,1)$.

The most systematic approach to this problem is to expand elements of K' as formal Laurent series in the local parameter $x-1$. We can do this using undeter-

[1] We are skipping some details here that will be covered in chapter 4.

mined coefficients as follows. Let $y := \sum_i a_i(x-1)^i$. Then $y \equiv a_0 \mod (x-1)$, so there are two choices for a_0, $+1$ or -1. Taking $a_0 = +1$, we get

$$\begin{aligned} y^2 - 1 &= 2a_1(x-1) + (2a_2 + a_1^2)(x-1)^2 + (2a_3 + 2a_1a_2)(x-1)^3 + \dots \\ &= (x-1)(x^2+x+2) \\ &= (x-1)((x-1)^2 + 3(x-1) + 4) \\ &= 4(x-1) + 3(x-1)^2 + (x-1)^3. \end{aligned}$$

From this, we obtain equations

$$\begin{aligned} 2a_1 &= 4, \\ 2a_2 + a_1^2 &= 3, \\ 2a_3 + 2a_1a_2 &= 1, \\ 2a_4 + 2a_1a_3 + a_2^2 &= 0, \\ &\vdots \end{aligned}$$

which can be successively solved for the coefficients a_i. Thus,

$$y = 1 + 2(x-1) - \frac{1}{2}(x-1)^2 + \frac{3}{2}(x-1)^3 + \dots$$

Now to expand $u = a(x) + b(x)y$ we just expand the rational functions $a(x)$ and $b(x)$ in powers of $x-1$, multiply $b(x)$ by y and combine terms. If all negative powers cancel and the constant terms do not, u is a local unit.

This example serves as a direct introduction to our next topic.

1.2 Completions

Given a ring R and an ideal I of R, we define the *completion* of R at I, denoted $\hat{R}_I$, to be the inverse limit $\varprojlim_n R/I^n$. Formally, $\hat{R}_I$ is the subring of the direct product

$$\prod_{n=1}^{\infty} R/I^n$$

consisting of those tuples $(r_1 + I, r_2 + I^2, \dots)$ such that $r_{n+1} \equiv r_n \mod I^n$, with pointwise operations. The canonical projection maps of the direct product restrict to $\hat{R}_I$, giving maps $\pi_n : \hat{R}_I \to R/I^n$ such that all of the diagrams

$$\begin{array}{ccc} \hat{R}_I & & \\ {\scriptstyle \pi_{n+1}}\downarrow & \searrow^{\pi_n} & \\ R/I^{n+1} & \longrightarrow & R/I^n \end{array}$$

commute, where the horizontal map is the natural map.

The projections π_n satisfy the following universal property:

Lemma 1.2.1. *Given any ring S and maps $\phi_n : S \to R/I^n$ such that all diagrams*

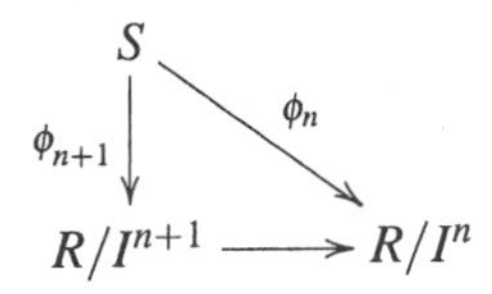

commute, there is a unique map $\phi : S \to \hat{R}_I$ making all diagrams

(1.2.2)

commute. □

We sometimes write $\phi = \overleftarrow{\lim}_n \phi_n$. In particular, there is a natural map $\phi : R \to \hat{R}_I$ whose kernel is easily seen to be $\cap_n I^n$. We say that R is *complete at I* when ϕ is an isomorphism.

Lemma 1.2.3. *A ring R is complete at the ideal I if and only if the following two conditions are satisfied:*

1. $\cap_{n=0}^{\infty} I^n = 0$, *and*
2. *Given any sequence $r_n \in R$ with $r_n \equiv r_{n+1} \mod I^n$ for all n, there exists $r \in R$ with $r \equiv r_n \mod I^n$ for all n.*

In particular, if $I^n = 0$ for some n, then R is complete at I.

Proof. As already noted, 1) is equivalent to the injectivity of the natural map $R \to \hat{R}_I$ and one verifies easily that 2) is equivalent to its surjectivity. If $I^n = 0$ for some n, the sequences satisfying 2) are eventually constant and we can take $r = r_n$ for any sufficiently large n. □

We will call a sequence r_n satisfying 2) above a *strong Cauchy* sequence, and an element r with $r \equiv r_n \mod I^n$ for all n a *limit*. In the presence of 1) such a limit is unique, and we write $r = \lim_n r_n$. Both conditions can therefore be reformulated as saying that every strong Cauchy sequence has a unique limit. More generally, given any sequence $x_i \in R$, the statement $\lim_i x_i = x$ means that for any integer $n > 0$, $x - x_i$ is eventually in I^n. In effect, we have introduced a topology on the ring R. Without belaboring this point, we note that it is immediate that the operations of addition and multiplication are continuous.

The ring $\hat{R}_I$ comes equipped with canonical projection maps $\pi_n : \hat{R}_I \to R/I^n$. If we let $\hat{I}$ be the set of sequences $\{x_n\}$ in $\hat{R}_I$ for which $x_n \in I$ for all n, we see that $\hat{I}^n$ is just the set of tuples the direct product whose first n components are zero, and thus $\ker \pi_n = \hat{I}^n$. It follows that the completion of $\hat{R}_I$ at $\hat{I}$ satisifies the same universal property as $\hat{R}_I$ does, so they are isomorphic. Finally, it is obvious that if $r = (r_1 + I, \ldots, r_n + I^n, \ldots) \in \hat{R}_I$, then $r = \lim_n \phi(r_n)$, where $\phi : R \to \hat{R}_I$ is the canonical map. Summarizing all of this, we have

Lemma 1.2.4. *With the above notation, $\hat{R}_I$ is complete at $\hat{I}$, $\hat{I}^n = \ker \pi_n$, and the image of R under the canonical map is dense. If $x_n, y_n \in \hat{R}_I$ with $x := \lim_n x_n$ and $y := \lim_n y_n$, then $xy = \lim_n x_n y_n$ and $x + y = \lim_n x_n + y_n$.* □

If we let R be the integers with I a prime ideal (p), we get the *p-adic integers* $\hat{\mathbb{Z}}_p$. Of more direct interest is the case $R = k[X]$, $I = (X)$; which yields the ring of formal power series $k[[X]]$. We will discuss this case further below.

Lemma 1.2.5. *Suppose that S is a subring of R, I is an ideal of R, and J is an ideal of S contained in I. Then there is a natural map $\phi : \hat{S}_J \to \hat{R}_I$ making all diagrams*

$$\begin{array}{ccc} \hat{S}_J & \xrightarrow{\phi} & \hat{R}_I \\ \downarrow \pi_n & & \downarrow \pi_n \\ S/J^n & \xrightarrow{\phi_n} & R/I^n \end{array}$$

commutative, where ϕ_n is induced by inclusion. If $R = S$, then ϕ is surjective. If, for almost all integers $n > 0$, there exists an integer m depending on n such that $S \cap I^m \subseteq J^n$, then ϕ is injective.

Proof. Since $J^n \subseteq I^n$ for any n, there are natural maps

$$\hat{S}_J \xrightarrow{\pi_n} S/J^n \xrightarrow{\phi_n} R/I^n$$

that commute with $R/I^{n+1} \to R/I^n$, so $\phi := \overleftarrow{\lim}_n(\phi_n \circ \pi_n)$ is defined, making the above diagrams commutative. From the definitions, we see that ϕ is surjective when $R = S$, and that $\ker \phi$ consists of those sequences $x = (x_n + J^n) \in \hat{S}_J$ with $x_n \in \ker \phi_n = S \cap I^n$ for all n. Choose such a sequence x and an integer n, and assume that there is an integer m, which we may take greater than n, with $S \cap I^m \subseteq J^n$. Since $x_m \in S \cap I^m \subseteq J^n$ and $x_m \equiv x_n \mod J^n$, we have $x_n \in J^n$ and thus $x = 0$ as required. □

We note for future reference that the notion of completeness for rings generalizes easily to modules. Suppose I is an ideal of R and M is an R-module. A strong Cauchy sequence in M is a sequence $\{x_n\}$ of elements of M such that $x_n \equiv x_{n+1} \mod I^n M$ for all n. We say that M is complete at I if every strong Cauchy sequence in M has a unique limit.

We turn now to the proof of Hensel's Lemma, which is the main result we need from the study of complete rings. We begin with a special case.

Lemma 1.2.6. *If R is complete at I and $u \in R$ is invertible modulo I, then u is invertible.*

Proof. By hypothesis there is an element $y \in R$ with $a = 1 - uy \in I$. Put $s_n := 1 + a + a^2 + \cdots + a^n$. Then $\{s_n\}$ is a strong Cauchy sequence, which therefore converges to some element $s \in R$. Since $(1-a)s_n = 1-a^{n+1}$, we obtain $(1-a)s = 1$ and thus $u^{-1} = ys$. $\square$

We have proved that if the polynomial $uX-1$ has a root mod I, then it has a root. Our main motivation for considering completions is to generalize this statement to a large class of polynomials.

Lemma 1.2.7 (Newton's Algorithm). *Let R be a ring with an ideal I and suppose that for some polynomial $f \in R[X]$ there exists $a \in R$ such that $f(a) \equiv 0 \mod I$ and $f'(a)$ is invertible, where $f'(X)$ denotes the formal derivative. Put*

$$b := a - \frac{f(a)}{f'(a)}.$$

Then $a \equiv b \mod I$ and $f(b) \equiv 0 \mod I^2$.

Proof. We have $b \equiv a \mod I$ because $f(a) \in I$. For any element $a \in R$ and any $n \geq 0$ we have the identity

$$X^n = (X-a+a)^n = \sum_{i=0}^{n} \binom{n}{i} (X-a)^i a^{n-i} = a^n + na^{n-1}(X-a) + h_n(X)(X-a)^2,$$

for some $h_n(X) \in R[X]$, whence

$$f(X) = f(a) + f'(a)(X-a) + h(X)(X-a)^2 \tag{1.2.8}$$

for some $h(X) \in R[X]$. With $X = b$ we have

$$f(b) = h(b)(b-a)^2 \in I^2.$$

$\square$

Newton's algorithm is quite effective computationally, because it converges very quickly. Using it, we obtain

Corollary 1.2.9 (Hensel's Lemma). *Let R be complete at an ideal I and let $f(X) \in R[X]$. Suppose, for some $u \in R$, that $f(u) \equiv 0 \mod I$ and that $f'(u)$ is invertible modulo I. Then there exists a unique element $v \in R$ satisfying $v \equiv u \mod I$ and $f(v) = 0$.*

Proof. By (1.2.6) every element of R congruent to $f'(u) \mod I$ is invertible. Put $u_1 = u$ and apply (1.2.7) to obtain an element $u_2 \equiv u_1 \mod I$ with $f(u_2) \equiv 0 \mod I^2$. Then $f'(u_2) \equiv f'(u_1)$, and therefore $f'(u_2)$ is invertible by the above remark.

This means that Newton's algorithm can be applied repeatedly to yield a strong Cauchy sequence $u = u_1, u_2, \ldots$ of elements of R such that $f(u_n) \in I^{2^{n-1}} \subseteq I^n$ for $n = 1, 2, \ldots$. By (1.2.3) the sequence has a limit $v \in R$. By continuity of addition and multiplication, we get $f(v) = \lim_n f(u_n) = 0$.

Some care is needed to prove uniqueness, because we are not assuming that R is an integral domain. Using (1.2.8) we have $f(X) = g(X)(X-v)$ for some polynomial $g(X) \in R[X]$. Then $f'(X) = g'(X)(X-v) + g(X)$. If v' is any root of f congruent to u modulo I, then

$$f'(u) \equiv f'(v') = g'(v')(v'-v) + g(v') \equiv g(v') \mod I,$$

whence $g(v')$ is a unit in R. But we have $0 = f(v') = g(v')(v-v')$, so $v = v'$ as required. □

We now apply these ideas to a discrete valuation ring $\mathcal{O}$ with maximal ideal P. Since $\cap_n P^n = 0$, the natural map $\mathcal{O} \to \hat{\mathcal{O}}_P$ is an embedding. We usually identify $\mathcal{O}_P$ with its canonical image in $\hat{\mathcal{O}}_P$. An important point is that v_P extends naturally to a discrete valuation on $\hat{\mathcal{O}}_P$.

Theorem 1.2.10. *Let $\mathcal{O}$ be a discrete valuation ring with maximal ideal P and local parameter t. Then $\hat{\mathcal{O}}_P$ is also a discrete valuation ring with local parameter t, and the natural projection $\pi_1 : \hat{\mathcal{O}}_P \to \mathcal{O}/P$ induces an isomorphism of residue fields $\hat{\mathcal{O}}_P/\hat{P} \simeq \mathcal{O}/P$, where $\hat{P}$ is the completion of P in $\hat{\mathcal{O}}_P$.*

Proof. We have $\hat{P} = \ker \pi_1$ by (1.2.4), so π_1 induces an isomorphism $\hat{\mathcal{O}}_P/\hat{P} \simeq \mathcal{O}_P/P$. Now (1.2.6) implies that every element of $\hat{\mathcal{O}}_P \setminus \hat{P}$ is a unit.

Since $P = \mathcal{O}_P t$ we have $\hat{P} = \hat{\mathcal{O}}_P t$, and because $\hat{\mathcal{O}}_P$ is complete at $\hat{P}$ by (1.2.4), it follows that no nonzero element of $\hat{P}$ is divisible by arbitrarily high powers of t. Thus, every element $x \in \hat{P}$ can written $x = ut^i$ for some $u \in \hat{\mathcal{O}}_P \setminus \hat{P}$. Since every such u is a unit, $\hat{\mathcal{O}}_P$ is an integral domain and thus is a is a discrete valuation ring with local parameter t. □

If K is the field of fractions of $\mathcal{O}$, we denote by $\hat{K}_P$ the field of fractions of $\hat{\mathcal{O}}_P$. We say that v_P is a complete discrete valuation of $\hat{K}_P$. If the natural map $K \to \hat{K}_P$ is an isomorphism, we say that K is complete at P. The embedding $\mathcal{O} \hookrightarrow \hat{\mathcal{O}}_P$ obviously extends to an embedding $K \hookrightarrow \hat{K}_P$.

Theorem 1.2.11. *Suppose that $\mathcal{O}_P$ is a discrete valuation ring of a field K, that K' is a finite extension of K, and that $\mathcal{O}_Q$ is an extension of $\mathcal{O}_P$ to K'. Let $e := e(Q|P)$ and $f := f(Q|P)$. Then there is a natural embedding $\hat{K}_P \to \hat{K}'_Q$, and if we identify $\hat{K}_P$ with its image in $\hat{K}'_Q$, then $e(\hat{Q}|\hat{P}) = e$, $f(\hat{Q}|\hat{P}) = f$, and $\hat{\mathcal{O}}_Q$ is a free $\hat{\mathcal{O}}_P$-module of rank ef generated by elements of $\mathcal{O}_Q$. In particular, $\hat{K}'_Q = K'\hat{K}_P$ and $|\hat{K}'_Q : \hat{K}_P| = ef$.*

Proof. From (1.2.5) we obtain a natural embedding $\hat{\mathcal{O}}_P \hookrightarrow \hat{\mathcal{O}}_Q$ that extends to $\hat{K}_P \hookrightarrow \hat{K}'_Q$. Choose local parameters t at P and s at Q. By (1.2.10), s and t are local parameters at $\hat{Q}$ and $\hat{P}$, respectively, and since $t = s^e u$ for some unit $u \in \mathcal{O}_Q \subseteq \hat{\mathcal{O}}_Q$, we have $e(\hat{Q}|\hat{P}) = e$. Using (1.2.5) and the natural isomorphisms of residue fields $\hat{\mathcal{O}}_Q/\hat{Q} \simeq F_Q$ and $\hat{\mathcal{O}}_P/\hat{P} \simeq F_P$ provided by (1.2.10), we get $f(\hat{Q}|\hat{P}) = f$.

In particular, $\hat{P}\hat{\mathcal{O}}_Q = \hat{Q}^e$, and $\dim_{F_P}(\hat{\mathcal{O}}_Q/\hat{P}\hat{\mathcal{O}}_Q) = ef$. Choose an F_P-basis $u_1, \ldots, u_{ef}$ for $\hat{\mathcal{O}}_Q/\hat{P}\hat{\mathcal{O}}_Q$. The u_i can be chosen to lie in $\mathcal{O}_Q$ because $\hat{\mathcal{O}}_Q = \mathcal{O}_Q + \hat{Q}^e$

for any e by (1.2.2). Moreover, the u_i are linearly independent over $\hat{\mathcal{O}}_P$, because given any nontrivial dependence relation we could divide by a power of t if necessary so that not all coefficients were divisible by t and obtain a nontrivial dependence relation modulo $\hat{P}$.

Let M be the free $\hat{\mathcal{O}}_P$-module generated by the u_i. It is clear that none of the u_i lie in $\hat{P}M$, which means that $\hat{P}^n M = \sum_i \hat{P}^n u_i$. If $\{m_n\}$ is a strong Cauchy sequence in M, we can write

$$m_n = \sum_{i=1}^{ef} a_{in} u_i$$

with $a_{in} \in \hat{\mathcal{O}}_P$, and it is clear that the sequence $\{a_{in}\}$ is a strong Cauchy sequence in $\hat{\mathcal{O}}_P$ for each i. Let $a_i = \lim_n a_{in}$ for each i. Then

$$\lim_n m_n = \sum_{i=1}^{ef} a_i u_i.$$

Since $\cap_n \hat{P}^n M \subseteq \cap_n \hat{Q}^n = 0$, the limit is unique, and therefore M is complete.

We have $\hat{\mathcal{O}}_Q = M + \hat{P}\hat{\mathcal{O}}_Q$, and we claim that in fact, $\hat{\mathcal{O}}_Q = M$.[2] Let $x \in \hat{\mathcal{O}}_Q$. Then $x = m_0 + x_0$ for some $m_0 \in M$ and $x_0 \in \hat{P}\hat{\mathcal{O}}_Q$. Inductively, assume that we have found elements $m_n \in M$ and $x_n \in \hat{P}^{n+1}\hat{\mathcal{O}}_Q$ with $m_{n-1} \equiv m_n \mod \hat{P}^n \hat{\mathcal{O}}_Q$ and $x = m_n + x_n$. Write

$$x_n = \sum_j a_j y_j$$

with $a_j \in \hat{P}^{n+1}$ and $y_j \in \hat{\mathcal{O}}_Q$. Then $y_j = m'_j + y'_j$ for some $m'_j \in M$ and $y'_j \in \hat{P}\hat{\mathcal{O}}_Q$. Put

$$m_{n+1} := m_n + \sum_j a_j m'_j \quad \text{and} \quad x'_{n+1} := \sum_j a_j y'_j.$$

Then the inductive step is easily verified, and we have constructed a strong Cauchy sequence $\{m_n\}$ in M, which therefore converges to some limit $m \in M$. But now we have $x - m \in \cap_n \hat{P}^n \hat{\mathcal{O}}_Q = 0$, and therefore $\hat{\mathcal{O}}_Q = M$ is a free $\hat{\mathcal{O}}_P$-module of rank ef with a basis contained in $\mathcal{O}_Q$. Extending scalars to $\hat{K}_P$, we obtain $|\hat{K'}_Q : \hat{K}_P| = ef$ and $\hat{K'}_Q = K'\hat{K}_P$, as claimed. □

We now specialize the discussion to the case of a complete discrete k-valuation ring for some ground field k, such that the residue field is a finite extension of the ground field.

Lemma 1.2.12. *Suppose that the k-algebra $\mathcal{O}$ is a complete discrete k-valuation ring with residue class map $\eta : \mathcal{O} \twoheadrightarrow F$. Assume further that F is a finite extension of k. Let F^{sep}/k be the maximal separable subextension of k. Then there is a unique k-algebra map $\mu : F^{\text{sep}} \to \mathcal{O}$ with $\eta \circ \mu = 1_{F^{\text{sep}}}$.*

---

[2]This would follow from Nakayama's lemma if we already knew that $\hat{\mathcal{O}}_Q$ was finitely generated.

Proof. Let $F^{\text{sep}} = k(u)$, where u is a root of the separable irreducible polynomial $f(X) \in k[X]$ and $\deg(f) = n$. Since $f(X)$ is separable, we have $f'(u) \neq 0$, so Hensel's Lemma (1.2.9) yields a unique root v of f in $\mathcal{O}$ with $\eta(v) = u$. Now, given any element $w \in F^{\text{sep}}$, there are uniquely determined elements $a_i \in k$ such that

$$w = \sum_{i=0}^{n-1} a_i u^i.$$

We define $\mu(w) := \sum_i a_i v^i \in \mathcal{O}$, and we easily check that μ splits the residue map. Because v is the unique root of f in $\mathcal{O}$ with residue u, it follows that μ is unique. □

Recall that the ring of formal power series $R[[X]]$ over some coefficient ring R is just the set of all sequences $\{a_0, a_1, \ldots\}$ of elements of R with elementwise addition, and with multiplication defined by $\{a_i\}\{b_j\} = \{c_k\}$, where

$$c_k = \sum_{i+j=k} a_i b_j.$$

Note that the sum is finite. We usually write the sequences as power series in some indeterminate:

$$f(X) = \sum_{i=0}^{\infty} a_i X^i,$$

but since the series is never evaluated at a nonzero element of R, the usual question of convergence does not arise. Nevertheless, the series is in fact a limit of its partial sums in a sense that we will make precise below.

Note that the formal derivative is a well-defined derivation, just as in the polynomial ring. Moreover, if R is an integral domain with field of fractions F, then $R[[X]]$ is an integral domain whose field of fractions is the field of formal Laurent series with coefficients in F of the form

$$f(X) = \sum_{i=-n}^{\infty} a_i X^i.$$

The field of formal Laurent series over F is denoted $F((X))$.

Lemma 1.2.13. *Let F be a field. Then $F[[X]]$ is a complete discrete valuation ring with local parameter X and residue field F.*

Proof. Define $v(\sum_i a_i X^i) = n$ if $a_i = 0$ for $i < n$ and $a_n \neq 0$. It is trivial to verify that v is a discrete valuation, so $F[[X]]$ is a valuation ring with maximal ideal M consisting of those power series with zero constant term. Let

$$f_n := \sum_{i=0}^{\infty} a_{ni} X^i$$

be a sequence of power series with $f_{n+1} \equiv f_n \mod M^n$. Then $a_{mn} = a_{nn}$ for all $m \geq n$, so $\{f_n\}$ converges to

$$f := \sum_{i=0}^{\infty} a_{ii} X^i. \quad \square$$

Note that the sequence of partial sums of a formal power series in $F[[X]]$ is a strong Cauchy sequence that converges to the infinite sum. More generally, in any complete ring we use the notation

$$x = \sum_{n=0}^{\infty} x_n$$

to indicate that the sequence of partial sums converges to x.

Theorem 1.2.14. *Suppose that the k-algebra $\mathcal{O}$ is a complete discrete k-valuation ring with residue class map $\eta : \mathcal{O} \twoheadrightarrow F$. Assume further that F is a finite separable extension of k. Given any local parameter t, there is a unique isometric isomorphism $\hat{\mu} : F[[X]] \simeq \mathcal{O}$ such that $\hat{\mu}(X) = t$.*

Proof. Let $\eta : \mathcal{O} \to F$ be the residue class map, and let $\mu : F \to \mathcal{O}$ be the unique splitting given by (1.2.12). Define $\hat{\mu} : F[[X]] \to \mathcal{O}$ via

$$\hat{\mu}\left(\sum_i a_i X^i\right) := \sum_i \mu(a_i) t^i.$$

This map is clearly well-defined and injective, and is uniquely determined by μ and t. To show that it is surjective, put $F' := \operatorname{im}(\mu)$. Then $\mathcal{O} = F' + P$, and $F' \cap P = 0$. Thus, for any $x \in \mathcal{O}$ there exists a unique element $a_0 \in F'$ with $x \equiv a_0 \mod P$. Choose a local parameter $t \in P$. Then there exists a unique $r_1 \in \mathcal{O}$ such that $x = a_0 + r_1 t$. An easy induction now shows that for any integer n there exist uniquely determined elements $a_0, \ldots, a_n \in F'$ and a uniquely determined element $r_{n+1} \in \mathcal{O}$ such that

$$x = \sum_{i=0}^{n} a_i t^i + r_{n+1} t^{n+1}.$$

Put $x_n := \sum_{i=0}^{n} a_i t^i$. Then $\lim_n x_n = x$. It follows that $x = \sum_{i=0}^{\infty} a_i t^i \in \operatorname{im}(\hat{\mu})$. $\square$

Corollary 1.2.15. *For every power series*

$$s = \sum_{n=1}^{\infty} a_n t^n \in k[[t]]$$

with $a_1 \neq 0$, there is a unique automorphism ϕ_s of $k[[t]]$ that is the identity on k and maps t to s.

Proof. This is immediate from (1.2.14) because s is a local parameter. $\square$

1.3 Differential Forms

Let R be a ring and M an R-module. A *derivation* of R into M is a map $\delta : R \to M$ such that

$$\delta(x+y) = \delta(x) + \delta(y),$$
$$\delta(xy) = x\delta(y) + \delta(x)y$$

for all $x, y \in R$. A standard example with $R = M = k[X]$ for some coefficient ring k (which is frequently a field) is the formal derivative:

$$\Big(\sum_i a_i x^i\Big)' = \sum_i i a_i x^{i-1}.$$

Notice that if we compose a derivation $\delta : R \to M$ with a homomorphism of R-modules $\phi : M \to N$, we get another derivation $\phi \circ \delta$. This suggests that there might be a universal derivation, from which all others can be obtained by composition in this way. In fact, we will make a slightly more general construction, as follows.

Let K be a k-algebra over some commutative ring k. By a k-derivation we mean a derivation δ that vanishes on $k \cdot 1$. By the product rule, this is equivalent to the condition that δ is k-linear. There is no loss of generality here, because we can take $k = \mathbb{Z}$ if we wish.

Observe that $K \otimes_k K$ is a K-module via $x(y \otimes z) = xy \otimes z$, and let D be the K-submodule generated by all elements of the form $x \otimes yz - xy \otimes z - xz \otimes y$. We define the K-module

$$\Omega_{K/k} := K \otimes_k K/D. \tag{1.3.1}$$

The relations D force the map $d_{K/k} : K \to \Omega_{K/k}$ given by

$$d_{K/k}(x) = 1 \otimes x + D$$

to be a k-derivation. We write $dx := 1 \otimes x + D$. Then $x \otimes y + D = xdy$. The map $d_{K/k}$ is in fact the universal k-derivation, namely we have

Theorem 1.3.2. *Let K be a k-algebra over a commutative ring k, M an K-module, and $\delta : K \to M$ a k-derivation. Then there exists a unique homomorphism ϕ : $\Omega_{K/k} \to M$ with $\delta = \phi \circ d_{K/k}$.*

Proof. Let $\phi'(x,y) = x\delta(y)$. Then ϕ' is k-bilinear so it factors uniquely through $K \otimes_k K$ by the universal property of tensor products. From the product rule, $\phi'(D) = 0$ and the rest is obvious. □

The elements of $\Omega_{K/k}$ are called *differential forms*, or sometimes *Kähler differentials*. Using (1.3.2), we can now naturally identify k-Derivations $\delta : K \to K$ (a common case) with elements of the dual $\mathrm{Hom}_K(\Omega_{K/k}, K)$. The standard case for us will be that K is a k-algebra over some ground field k that we are thinking of as "constants" and all derivations will be k-derivations. When there is no danger of confusion, we may conserve notation by dropping the subscript and writing $d := d_{K/k}$. Sometimes, however, we may need to retain the subscript K and write $d_K := d_{K/k}$.

Note that the set $\{dx \mid x \in K\}$ generates Ω_K as an K-module, but is not in general equal to all of Ω_K. Differential forms that happen to be of the form dx for some $x \in K$ are called *exact*. The exact differentials form a k-subspace of Ω_K.

The following functorial properties of the differential map are useful.

Lemma 1.3.3. *Suppose $\phi : K \to K'$ is a k-algebra map. Then there exists a unique map $d\phi$ making the diagram*

$$\begin{array}{ccc} K & \xrightarrow{\phi} & K' \\ \downarrow{d_K} & & \downarrow{d_{K'}} \\ \Omega_K & \xrightarrow{d\phi} & \Omega_{K'} \end{array}$$

commute. Moreover, given another k-algebra map $\phi' : K' \to K''$ we have

$$d(\phi' \circ \phi) = (d\phi') \circ (d\phi).$$

Proof. The composition $d_{K'} \circ \phi : K \to \Omega_{K'}$ is certainly a derivation, so there is a unique map $d\phi : \Omega_K \to \Omega_{K'}$ making the above diagram commute. It is easy to check that

$$\begin{array}{ccc} K & \xrightarrow{\phi' \circ \phi} & K'' \\ \downarrow{d_K} & & \downarrow{d_{K''}} \\ \Omega_K & \xrightarrow{(d\phi') \circ (d\phi)} & \Omega_{K''} \end{array}$$

is commutative, so uniqueness yields $d(\phi' \circ \phi) = (d\phi') \circ (d\phi)$, as required. □

Suppose now that $K \subseteq K_1$ are k-algebras, M is a K_1-module, and δ is a k-derivation of K into M. We ask whether δ is the restriction of a k-derivation of K_1 into M. This question can be converted into a problem of extending homomorphisms instead of derivations by means of the following construction.

Put $A := K_1 \oplus M$ (vector space direct sum). Then the product $(x_1 + m_1)(x_2 + m_2) := x_1x_2 + x_1m_2 + x_2m_1$ converts A to a k-algebra such that the projection $\pi : A \to K_1$ is a homomorphism. It is straightforward to verify that the map $D : K \to A$ given by $D(x) = x + \delta(x)$ is a k-algebra homomorphism, and that δ extends to a derivation $\delta_1 : K_1 \to M$ if and only if D extends to a homomorphism $D_1(x) = x + \delta_1(x)$ of K_1 into A.

Note that A is actually a graded k-algebra; that is, there is a direct sum decomposition

$$A = \bigoplus_{i \geq 0} A_i$$

with $A_iA_j \subseteq A_{i+j}$. Every such algebra has an ideal

$$M := \bigoplus_{i>0} A_i,$$

and we say that A is *complete* if it is complete with respect to M (see Section 1.2). In the above case, we have $A_i = 0$ for $i > 1$, so $M^2 = 0$ and A is complete by (1.2.3).

Given a map $D : K \to A$, let $D^{(i)}$ denote the composition with projection onto A_i. Using Hensel's lemma, we can further reduce the extension problem to the problem of extending $D^{(0)}$ when $K \subseteq K_1$ is a finite separable extension of fields.

Theorem 1.3.4. *Suppose that $k \subseteq K \subseteq K_1$ are fields, K_1/K is finite and separable, A is complete graded k-algebra, and $D : K \to A$ is a k-algebra homomorphism. Given any extension $D_1^{(0)}$ of $D^{(0)}$ to K_1, there exists a unique extension D_1 of D to K_1 such that the diagram*

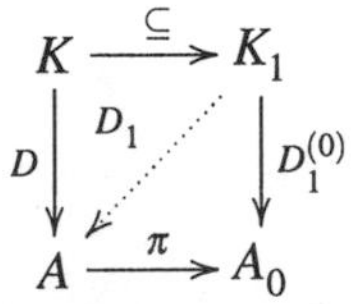

$$\begin{array}{ccc} K & \xrightarrow{\subseteq} & K_1 \\ {\scriptstyle D}\downarrow & \overset{D_1}{\swarrow} & \downarrow{\scriptstyle D_1^{(0)}} \\ A & \xrightarrow{\pi} & A_0 \end{array}$$

is commutative.

Proof. By (A.0.17) we have $K_1 = K(u)$ for some element u with separable minimum polynomial $f(X) \in K[X]$. Put $v := D_1^{(0)}(u) \in A_0$. Then v is a root of $f_1 := D_1^{(0)}(f) = D^{(0)}(f) \in A_0[X]$. Furthermore, $f_1'(v)$ is invertible in A_0 because $f'(u)$ is invertible in K_1. Now consider the polynomial $D(f) \in A[X]$. We have $D(f) \equiv f_1 \mod M$ where M is the maximal graded ideal of A. It follows that v is a root of $D(f)$ modulo M and that $D(f)'(v)$ is invertible modulo M. By Hensel's Lemma, there is a unique root v_1 of $D(f)$ in A congruent to v modulo M. To each such root there corresponds a unique extension D_1 of D to K_1, defined by $D_1(u) = v_1$. □

Applying the theorem to the k-algebra $K_1 \oplus M$ as described above, we obtain

Corollary 1.3.5. *Let K_1/K be a finite extension of fields. Then K_1/K is separable if and only if every derivation of K into a K_1-module M extends uniquely to K_1.*

Proof. One implication is immediate from the theorem. Conversely, if K_1/K is inseparable, there is a subfield $K \subseteq E \subseteq K_1$ where K_1/E is purely inseparable of degree $p = \mathrm{char}(K)$ (see (A.0.9)). Thus, we have $K_1 \simeq E[X]/(X^p - a)$. Since the formal derivative on $E[X]$ vanishes at $X^p - a$, it induces a nonzero derivation on K_1 that vanishes on E and therefore on K. □

We want to apply (1.3.5) to the special case that K is a finite, separable extension of $k(x)$ for some $x \in K$ transcendental over a subfield k. In this situation, we say that x is a *separating variable* for K/k. In particular, we have $\mathrm{trdeg}(K/k) = 1$.

Corollary 1.3.6. *Suppose that $k \subseteq K$ are fields and x is a separating variable for K/k. Then $\dim_K \Omega_{K/k} = 1$ and $d_{K/k}(x) \neq 0$.*

Proof. If $K = k(x)$, the formal derivative is a nonzero derivation, so $dx \neq 0$. From the sum, product, and quotient rules, every derivation on $k(x)$ is determined by its value at x, so the universal property of dx implies that dx is a $k(x)$-basis for $\Omega_{k(x)/k}$.

In general, (1.3.5) implies that the natural map $\Omega_{k(x)} \longrightarrow \Omega_K$ is nonzero, and that the image of dx in Ω_K is a basis. □

Note that if $x \in K$ is a separating variable, then for every $y \in K$ we have

$$dy = \frac{dy}{dx} dx$$

for some well-defined function $dy/dx \in K$ because $\dim_K \Omega_K = 1$.

There are further consequences to be obtained from (1.3.4). Suppose R is a k-algebra and we put $A := R[[t]]$, the k-algebra of formal power series with coefficients in R. Since A is a complete graded k-algebra, the theorem tells us that a homomorphism $D : K \longrightarrow A$ can be extended to K_1, provided that the projection $D^{(0)} : K \longrightarrow R$ can be extended. What is this saying?

The map D is determined by the family of k-linear maps $\{D^{(n)} : K \rightarrow R \mid n = 0, 1, \ldots\}$ defined by

$$D(x) = \sum_{n=0}^{\infty} D^{(n)}(x) t^n$$

for $x \in K$. The condition that D is a homomorphism is equivalent to the following condition for each nonnegative integer n:

$$D^{(n)}(xy) = \sum_{i=0}^{n} D^{(i)}(x) D^{(n-i)}(y) \tag{1.3.7}$$

for all $x, y \in K$. In particular, for $n = 0$ (1.3.7) says that $D^{(0)}$ is a homomorphism, and, for $n = 1$, that if we convert R to a K-module via

$$x \cdot r := D^{(0)}(x) r$$

for $x \in K$ and $r \in R$, then $D^{(1)}$ is a derivation of K with coefficients in R. For this reason, we call the map D a *generalized derivation of K with coefficients in R.*

Corollary 1.3.8. *Suppose that K_1/K is a finite separable extension of fields over k, and that D is a generalized derivation of K with coefficients in some k-algebra R. For every extension $D_1^{(0)}$ of $D^{(0)}$ to K_1, there exists a unique extension D_1 of D to K_1.* □

Even though D is a map to $R[[t]]$, we abuse notation by writing $D : K \longrightarrow R$ because we are thinking of D as a family of maps $D^{(n)} : K \longrightarrow R$.

The standard application of (1.3.8) is to the formal derivatives on $k[X]$. Define, for nonnegative integers m,n,

$$D^{(n)}(X^m) = \begin{cases} \binom{m}{n}X^{m-n} & \text{if } m \geq n, \\ 0 & \text{otherwise,} \end{cases}$$

and extend linearly to $k[X]$. These maps are readily verified to define a generalized derivation $D : k[X] \to k[X]$ which we will call the *Hasse derivative with respect to* X on $k[X]$. Note that $D^{(1)}$ is just the standard formal derivative. In characteristic zero, we have

$$D^{(n)} = \frac{1}{n!}\frac{d^n}{dX^n}$$

but in finite characteristic the Hasse derivative is more interesting. We first extend the Hasse derivative to $k(X)$ via

Lemma 1.3.9. *Suppose that R is an integral domain and $D : R \to A$ is a homomorphism for some complete graded algebra A. If $A^{(0)}$ is a field and $D^{(0)}$ is an embedding, then D extends uniquely to the field of fractions of R.*

Proof. Since $A^{(0)}$ is a field, every element of A with a nonzero component in degree zero is invertible by (1.2.6). Since we are assuming that $D^{(0)}(r) \neq 0$ for all nonzero $r \in R$, D extends uniquely to the field of fractions. $\square$

In particular, the Hasse derivative extends uniquely to a generalized derivation of $k(X)$ into $k(X)$. For example, from the product rule, we have, for $n \geq 1$,

$$0 = D^{(n)}(1) = D^{(n)}(XX^{-1}) = \sum_{i=0}^{n} D^{(i)}(X)D^{(n-i)}(X^{-1}) = XD^{(n)}(X^{-1}) + D^{(n-1)}(X^{-1}),$$

whence a simple induction yields $D^{(n)}(X^{-1}) = (-1)^n X^{-n-1}$. By a slightly more elaborate induction, we obtain

$$D^{(n)}(X^{-i}) = (-1)^n \binom{n+i-1}{n} X^{-n-i} \tag{1.3.10}$$

for all positive integers n,i. Finally, using (1.3.8), we have

Theorem 1.3.11. *Suppose that $x \in K$ is a separating variable for K/k and that K' is any field containing K. Then the Hasse derivative on $k[x]$ extends uniquely to a generalized derivation $D_x : K \to K'$. In particular, D_x has coefficients in K.* $\square$

We continue to call the extended map $D_x^{(n)}$ the nth Hasse derivative with respect to x. However, (1.3.6) yields a map $\delta_x : K \to K$ given by

$$dy = \delta_x(y)dx,$$

which is easily seen to be a derivation. Since $\delta_x(x) = 1 = D_x^{(1)}(x)$, it follows that δ_x and $D_x^{(1)}$ are both extensions of the formal derivative on $k(x)$. From the uniqueness of such an extension given by (1.3.5) we obtain

Lemma 1.3.12. *Let $x \in K$ be a separating variable. Then for any $y \in K$ we have*

$$D_x^{(1)}(y) = \frac{dy}{dx}. \quad \square$$

The higher Hasse derivatives are closely related to ordinary higher derivatives. Indeed, if we let D_x^n denote the iterated derivative $d^n/dx^n = D_x^{(1)} \circ D_x^{(1)} \circ \cdots \circ D_x^{(1)}$, then a simple induction on n shows that

$$D_x^n(yz) = \sum_{i=0}^{n} \binom{n}{i} D_x^i(y) D_x^{n-i}(z).$$

If we now let $A := K[[t]]/(t^p)$ where $p := \mathrm{char}(K)$, then A is a complete graded k-algebra by (1.2.3), and the map

$$\tilde{D}_x(y) := \sum_{i=0}^{p-1} \frac{1}{i!} D_x^i(y) t^i$$

is a k-algebra homomorphism $K \longrightarrow A$. However, D_x also induces a k-algebra map $K \longrightarrow A$. If $\pi_0 : A \longrightarrow K$ is the projection onto the degree zero component, then $\pi_0 \circ D_x = \pi_0 \circ \tilde{D}_x = 1_K$. Since both maps have the same restriction to $k[x]$, (1.3.9) and (1.3.4) imply

Lemma 1.3.13. *If* $\mathrm{char}(K) = 0$ *or if* $n < \mathrm{char}(K)$ *and $x \in K$ is a separating variable, then*

$$D_x^{(n)} = \frac{1}{n!} D_x^n,$$

where D_x^n is the n-fold iterated first Hasse derivative. $\square$

In addition to the product rule, the Hasse derivatives also satisfy the chain rule. If we compose $D_x : K \longrightarrow K[[t]]$ with an automorphism ϕ of $K[[t]]$ that is the identity on $K[[t]]_0$, the result is another generalized derivation. Suppose that $y \in K$ is also a separating variable. Then $D_x^{(1)}(y) \neq 0$ by (1.3.6), whence (1.2.15) provides an automorphism ϕ of $K[[t]]$ that is the identity on $K[[t]]_0$ and that satisfies

$$\phi(t) = \sum_{i=1}^{\infty} D_x^{(i)}(y) t^i.$$

Then $(\phi^{-1} \circ D_x)(y) = y + t = D_y(y)$. Thus, $\phi^{-1} \circ D_x$ and D_y agree on $k[y]$, so they are equal by (1.3.11), and we have $D_x = \phi \circ D_y$. Explicit formulas for $D_x^{(n)}$ in terms of $D_y^{(i)}$ can be extracted from this, but they are rather messy. Fortunately, we only need one coefficient explictly for a later application.

Lemma 1.3.14. *Suppose that $k \subseteq K$ are fields and $x, y \in K$ are separating variables for K/k. Then there are functions $d_1, \ldots, d_{n-1} \in K$ that are polynomial expressions in $D_x^{(i)}(y)$ for $1 \leq i \leq n$, such that for any $f \in K$ we have*

$$D_x^{(n)}(f) = (dy/dx)^n D_y^{(n)}(f) + \sum_{i=1}^{n-1} d_i D_y^{(i)}(f).$$

Proof. Let ϕ be the automorphism of $K[[t]]$ described above, and put

$$s := \phi(t) = \sum_{i=1}^{\infty} D_x^{(i)}(y)t^i.$$

Then for any $f \in K$ we have

$$D_x(f) = \sum_{i=0}^{\infty} D_y^{(i)}(f)s^i.$$

In particular, $D_x^{(n)}(f)$ is the coefficient of t^n in the right-hand side. Let d_i be the coefficient of t^n in s^i for $1 \leq i \leq n$. Then $d_n = (D_x^{(1)}(y))^n = (dy/dx)^n$. □

Corollary 1.3.15. *With the notation of the lemma, suppose that* $\operatorname{char}(K) = p$ *and that* $f \in K^p$. *Then*

$$D_x^{(p)}(f) = (D_x^{(1)}(y))^p D_y^{(p)}(f).$$

Proof. By (1.3.13) we have $D_y^{(i)}(f) = 0$ for $0 < i < p$. □

1.4 Residues

In this section, we discuss Tate's elegant theory of abstract residues, closely following [20]. For a variation based on topological ideas, see the appendix of [13].

Let V be a (not necessarily finite-dimensional) vector space over a field k. Recall that a k-linear map $y : V \to V$ has *finite rank* if $y(V)$ is finite-dimensional. We can generalize this notion by calling y *finitepotent* if $y^n(V)$ is finite-dimensional for some positive integer n. Equivalently, there is a finite-dimensional y-invariant subspace $W \subseteq V$ such that y is nilpotent on V/W. We call such a subspace a *core subspace* for y. Denote by $\operatorname{tr}_W(y)$ the trace of $y|_W$.

Lemma 1.4.1. *Suppose* $y : V \to V$ *is* k*-linear and* W, U *are core subspaces. Then* $\operatorname{tr}_W(y) = \operatorname{tr}_U(y)$.

Proof. We may assume, without loss of generality, that $V = W + U$ is finite-dimensional. Then $\operatorname{tr}_V(y) = \operatorname{tr}_{V/W}(y) + \operatorname{tr}_W(y) = \operatorname{tr}_W(y)$ and similarly, $\operatorname{tr}_V(y) = \operatorname{tr}_U(y)$. □

We can therefore unambiguously define $\operatorname{tr}_V(y)$ for any finitepotent map y. The following result is easy.

Lemma 1.4.2. *Let* $y : V \to V$ *be finitepotent, and suppose that* $W \subseteq V$ *is* y*-invariant. Then* y *is finitepotent on* W *and* V/W *and*

$$\operatorname{tr}_V(y) = \operatorname{tr}_W(y) + \operatorname{tr}_{V/W}(y). \quad \square$$

Lemma 1.4.3. *If* y *and* x *are any two* k*-linear maps on* V *and* yx *is finitepotent, then* xy *is also finitepotent and* $\operatorname{tr}_V(yx) = \operatorname{tr}_V(xy)$.

Proof. If $W = (yx)^n(V)$ is finite-dimensional, then

$$U := (xy)^{n+1}(V) = x \circ (yx)^n \circ y(V) \subseteq x(W),$$

so U is also finite-dimensional. Moreover, by choosing n large enough, we may assume that $yx(W) = W$ and $xy(U) = U$. This implies that $y : U \to W$ and $x : W \to U$ are both isomorphisms, from which it follows that $\mathrm{tr}_U(xy) = \mathrm{tr}_W(yx)$. □

Call a k-subspace E of $\mathrm{End}_k(V)$ finitepotent if there exists an integer n such that for every word w of length n in the elements of E, $w(V)$ is finite-dimensional.

Lemma 1.4.4. *If E is a finitepotent subspace of $\mathrm{End}_k(V)$, then $\mathrm{tr} : E \to k$ is k-linear.*

Proof. Take $y, x \in E$ and for any nonnegative integer n, put

$$V_n := \sum_w w(V),$$

where the sum is taken over all words w of length n in x and y. If w_0 is any initial segment of w, then $w(V) \subseteq w_0(V)$, and in particular, $V_n \subseteq V_{n-1}$. This implies that V_n is invariant under y and x. For sufficiently large n, it follows that V_n is a core subspace for both x and y, and linearity of tr_V follows from linearity of tr_{V_n}. □

We note that some hypothesis such as the above is necessary in order to get additivity of the trace. See Exercise 1.15 for an interesting counterexample due to G. Bergman. It is clear, however, that any product of linear maps in which at least one factor has finite rank remains of finite rank. In particular, we have

Lemma 1.4.5. *If E is a finitepotent subspace and x has finite rank, then $\langle E, x\rangle$ is finitepotent.* □

Next, suppose that W, W' are subspaces of V. We say that W is *nearly contained* in W' and write $W \preceq W'$ if $\dim(W/W \cap W') < \infty$, and define $W \sim W'$ if $W \preceq W'$ and $W' \preceq W$. Then y is finitepotent if $y^n(V) \preceq 0$ for some n.

Note that if $\mathscr{O}$ is a discrete k-valuation ring of K whose residue field is a finite extension of k, then $x\mathscr{O} \preceq \mathscr{O}$ for all $x \in K$.

The following properties are straightforward consequences of the isomorphism theorems:

Lemma 1.4.6. *If $W \preceq W'$ and $y \in \mathrm{End}_k(V)$, then $y(W) \preceq y(W')$. If also $W' \preceq W''$, then $W \preceq W''$. In particular, $\sim$ is an equivalence relation. Moreover, if $W_i \preceq W_i'$ $(i = 1, 2)$, then $W_1 + W_2 \preceq W_1' + W_2'$.* □

Now for $W, W' \subseteq V$ define

$$E_V(W, W') := \{y \in \mathrm{End}_k(V) \mid y(W) \preceq W'\}.$$

Lemma 1.4.7. *$E_V(W, W')$ is a k-subspace of $\mathrm{End}_k(V)$. If $y \in E_V(W, W')$, $W' \preceq U$, and $x \in E_V(U, U')$, then $xy \in E_V(W, U')$. Moreover, if we put $E_1 := E_V(V, W)$, $E_2 := E_V(W, 0)$, and $E := E_V(W, W)$, then E_1 and E_2 are two-sided ideals of E, $E := E_1 + E_2$, and $E_0 := E_1 \cap E_2$ is finitepotent.*

Proof. Let $y,x \in E_V(W,W')$ and $\alpha \in k$. Then $(\alpha y+x)(W) \subseteq y(W)+x(W) \preceq W'$ by (1.4.6). Moreover, if $y \in E_V(W,W')$, $W' \preceq U$, and $x \in E_V(U,U')$, then $y(W) \preceq W'$, so $xy(W) \preceq x(W') \preceq x(U) \preceq U'$ by (1.4.6). In particular, E_1 and E_2 are two-sided ideals of E, and $E_0^2 \subseteq E_V(V,0)$ is finitepotent. Finally, let $\pi : V \to W$ be an arbitrary projection map, and let $y \in E$. Then $\pi y \in E_1$ and $(1-\pi)y \in E_2$, whence $E = E_1 + E_2$. □

Define the *near-stabilizer* of a chain $V = V_0 \supseteq V_1 \supseteq \cdots \supseteq V_n = 0$ to be the set

$$\bigcap_{i<n} E_V(V_i, V_{i+1}).$$

Corollary 1.4.8. *The near-stabilizer of a chain is a finitepotent subspace of* $\mathrm{End}_k(V)$. □

Let $y : V \to V$ be any k-linear map. We say that $W \subseteq V$ is *nearly y-invariant* if $y(W) \preceq W$. Consider now a k-algebra K and a K-module V with a k-subspace W that is nearly y-invariant for all $y \in K$. We will call such a subspace a *near submodule*. An element $y \in K$ induces a k-linear transformation in $E := E_V(W,W)$ that, by abuse of notation, we will continue to call y. Define E_1, E_2, and E_0 as above. Write $y = y_1 + y_2$ with $y_i \in E_i$. If $x \in K$ is another element and we also write $x = x_1 + x_2$ with $x_i \in E_i$, then the commutator is $[y,x] := yx - xy$, which is of course zero since K is commutative. Expanding the commutator, we have

$$(*) \qquad 0 = [y_1,x_1] + [y_1,x_2] + [y_2,x_1] + [y_2,x_2].$$

Note that y_1x_2 and x_2y_1 are both in $E_1 \cap E_2 = E_0$ since the E_i are ideals, so $[y_1,x_2] \in E_0$. Similarly, $[y_2,x_1] \in E_0$, so $(*)$ implies that $[y_1,x_1] \equiv -[y_2,x_2] \mod E_0$. However, $[y_i,x_i] \in E_i$ $(i = 1,2)$, so we conclude that $[y_i,x_i] \in E_0$ $(i = 1,2)$, and in particular, $\mathrm{tr}_V([y_1,x_1])$ is defined.

However, $y_1x_2 \in E_0$, so that $\mathrm{tr}_V(y_1x_2)$ is defined, and therefore $\mathrm{tr}_V([y_1,x_2]) = 0$ by (1.4.3). Similarly, $\mathrm{tr}_V([y_2,x_1]) = 0$. Since $[y,x_1] - [y_1,x_1] = [y-y_1,x_1] = [y_2,x_1]$, it follows that $\mathrm{tr}_V([y,x_1]) = \mathrm{tr}_V([y_1,x_1])$ is independent of the choice of decomposition $y = y_1 + y_2$, and similarly for x. If $\pi : V \to W$ is a projection, we may take $y_1 = \pi y$. Note that if W is actually invariant under y and x, then $[\pi y, x]$ actually stabilizes the chain $V \supseteq W \supseteq 0$ and is therefore nilpotent.

Finally, note that since $[\pi y, \pi x]$ nearly stabilizes $V \supseteq W \supseteq 0$, the finite-dimensional subspace $[\pi y, \pi x](W)$ is a core subspace for $[\pi y, \pi x]$. Summarizing this argument, we have obtained the following remarkable facts:

Lemma 1.4.9. *If $W \subseteq V$ is nearly invariant under commuting maps y,x, and $\pi : V \to W$ is any projection, then $[\pi y, \pi x]$ nearly stabilizes the chain $V \supseteq W \supseteq 0$, and $\mathrm{tr}_V[\pi y, \pi x]$ is independent of π. If W is actually invariant under y and x, then $\mathrm{tr}_V[\pi y, \pi x] = 0$. Moreover, if $W_0 := [\pi y, \pi x](W)$, then W_0 is finite-dimensional and*

$$\mathrm{tr}_V[\pi y, \pi x] = \mathrm{tr}_{W_0}[\pi y, \pi x]. \quad \square$$

Thus, we have unambiguously defined a function $K \times K \to k$:

$$\langle y,x \rangle_{V,W} := \mathrm{tr}_V[\pi y, \pi x] = \mathrm{tr}_V[\pi y, x] = \mathrm{tr}_V[x, \pi y],$$

which is easily seen to be an alternating k-bilinear form. We call this form the *residue form* afforded by the pair (V,W).

Lemma 1.4.10. *If V is a K-submodule of V', then $\langle y,x\rangle_{V',W} = \langle y,x\rangle_{V,W}$ for all $y,x \in K$. If $W' \subseteq V$ and $W' \sim W$, then W' is a near K-submodule and $\langle y,x\rangle_{V,W'} = \langle y,x\rangle_{V,W}$ for all $y,x \in K$.*

Proof. Since core subspaces for all finitepotent maps under consideration lie in W, enlarging V has no effect, and the first statement is immediate. The second easily reduces to the case that $W' \subseteq W$, since W and W' both have finite index in $W+W'$. If $\pi' : V \longrightarrow W'$ is a projection, we can write $\pi = \pi' + \pi''$, where $\pi : V \longrightarrow W$ and π'' is a projection onto a finite-dimensional complement to W' in W. Then $\pi'' yx$ has finite rank, so $\mathrm{tr}([\pi'' y,x]) = 0$ by (1.4.3) and the result follows from (1.4.4) and (1.4.5). □

Theorem 1.4.11. *If W_1 and W_2 are near K-submodules of V, then so are W_1+W_2 and $W_1 \cap W_2$, and*

$$\langle y,x\rangle_{V,W_1+W_2} - \langle y,x\rangle_{V,W_1} - \langle y,x\rangle_{V,W_2} + \langle y,x\rangle_{V,W_1\cap W_2} = 0$$

for all $y,x \in K$.

Proof. Let $y \in K$. Then $y(W_i) \preceq W_i$, so certainly $y(W_i) \preceq W_1+W_2$ for $i=1,2$. Thus $y(W_1+W_2) \subseteq y(W_1)+y(W_2) \preceq W_1+W_2$ by (1.4.6), and it follows that W_1+W_2 is a near submodule.

Let $\eta_i : V \longrightarrow V/W_i$ be the quotient map $(i=1,2)$ and let $U := y(W_1 \cap W_2)$. Then $\eta_i(U)$ is finite-dimensional for $i=1,2$. Hence $\eta_1 \oplus \eta_2(U)$ is also finite dimensional, and thus $W_1 \cap W_2$ is also a near submodule.

It remains to prove that the alternating sum is zero. Put $W_0 := W_1 \cap W_2$ and choose subspaces W_1', W_2', W_3 such that $W_i = W_0 \oplus W_i'$ $(i=1,2)$ and $V = W_3 \oplus (W_1+W_2)$. Then we have a direct sum decomposition

$$V = W_0 \oplus W_1' \oplus W_2' \oplus W_3$$

and a corresponding decomposition of the identity into four mutually orthogonal projection maps

$$1_V = \pi_0 + \pi_1' + \pi_2' + \pi_3.$$

Put $\pi_i := \pi_0 + \pi_i'$ for $i=1,2$ and let $y,x \in K$. Dropping the subscript V for now, we have

$$\langle y,x\rangle_{W_i} = \mathrm{tr}[\pi_i y,x], \quad (i=0,1,2), \text{ and}$$

$$(*) \qquad \langle y,x\rangle_{W_1+W_2} = \mathrm{tr}[\pi_1 y + \pi_2 y - \pi_0 y, x].$$

We want to expand the commutators in $(*)$ and use additivity of the trace. Before doing so, however, we need to verify that all commutators lie in some finitepotent subspace of $\mathrm{End}_k(V)$.

To this end, let E be the near-stabilizer of the chain $V \supseteq W_1+W_2 \supseteq W_1 \cap W_2 \supseteq 0$, which is finitepotent by (1.4.8). We argue that $[\pi_i y,x] \in E$ for $i=0,1,2$. This is obvious for $i = 0$. For $i = 1,2$ we clearly have

$$[\pi_i y, x] \in E(V, W_i) \cap E(W_i, 0) \subseteq E(V, W_1 + W_2) \cap E(W_0, 0).$$

The problem is to show that $[\pi_i y, x](W_1 + W_2) \preceq W_0$, which immediately reduces to

$$[\pi_i y, x](W_{3-i}) \preceq W_0.$$

However, this follows by observing that $\pi_i \pi_{3-i} = \pi_0$ and the fact that W_{3-i} is a near submodule.

Now we can expand $(*)$ and conclude that the alternating sum is zero, as required. □

We next need to provide a connection between the residue form and the module of differential forms. This is given by:

Theorem 1.4.12. *Let K be a k-algebra, V a K-module, and $W \subseteq V$ a near submodule. Then there is a k-linear function $\mathrm{Res}_W^V : \Omega_{K/k} \to k$ that vanishes on exact differential forms such that*

$$\mathrm{Res}_W^V(ydx) = \langle y, x \rangle_{V,W} \quad \textit{for all } y, x \in K.$$

Moreover, $\mathrm{Res}_W^V(x^n dx)$ vanishes for all $n \geq 0$, and if x is invertible, it vanishes for all $n \neq -1$. If $x(W) \subseteq W$, put $W_0 : W \cap y^{-1}(W)$. Then

$$(*) \qquad \mathrm{tr}_V[\pi y, x] = \mathrm{tr}_{W/W_0}([\pi y, x])$$

for any projection $\pi : V \to W$. Finally, if x is invertible and W is invariant under x and y, then

$$(1.4.13) \qquad \mathrm{Res}_W^V(yx^{-1}dx) = \mathrm{tr}_{W/x(W)}(y).$$

Proof. Let $y, x, w \in K$ and decompose each of them using $E_V(W, W) = E_1 + E_2$ as above. Then we have the following identities:

$$[y_1, x_1 w_1] = y_1 x_1 w_1 - x_1 w_1 y_1,$$
$$[y_1 x_1, w_1] = y_1 x_1 w_1 - w_1 y_1 x_1,$$
$$[y_1 w_1, x_1] = y_1 w_1 x_1 - x_1 y_1 w_1.$$

All three commutators lie in the same finitepotent subspace E_0, so the trace is linear. Subtracting the second and third equations from the first and taking the trace, we get

$$\begin{aligned} \langle y, xw \rangle - \langle yx, w \rangle - \langle yw, x \rangle &= \mathrm{tr}(w_1 y_1 x_1 + x_1 y_1 w_1 - x_1 w_1 y_1 - y_1 w_1 x_1) \\ &= \mathrm{tr}(x_1 [y_1, w_1] - [y_1, w_1] x_1) \\ &= 0 \end{aligned}$$

by (1.4.3). (Note that $[y_1, w_1] \in E_0$ and $x_1 \in E_1$.) Now the definition of $\Omega_{K/k}$ (1.3.1) implies that the residue form factors uniquely through $\Omega_{K/k}$ via a k-linear map $\mathrm{Res}_W^V : \Omega_{K/k} \to k$, as advertised.

To compute $\mathrm{Res}(x^n dx)$, write $x = x_1 + x_2$ with $x_i \in E_i$ as before, and note that for $n \geq 0$ we have $x^n = x_1^n + x_{2,n}$ for some $x_{2,n} \in E_2$ because the E_i are ideals. Hence, $\mathrm{Res}_W^V(x^n dx) = \mathrm{tr}_V([x_1^n, x_1]) = 0$. In particular, $\mathrm{Res}_W^V(dx) = 0$ for any exact differential. If x is invertible and $n \leq -2$, we have

$$\mathrm{Res}(x^n dx) = \mathrm{Res}\big(-(x^{-1})^{-n-2}\big)d(x^{-1}) = 0.$$

Suppose that $x(W) \subseteq W$ and that $\pi : V \to W$ is a projection. After verifying that $[\pi y, x]$ maps V into W and is zero on W_0, we have $(*)$.

Finally, if x is invertible, and W is invariant under x and y, we apply $(*)$ with yx^{-1} in place of y. Here, $x(W) \subseteq W_0$ and expanding the commutator, we conclude that

$$\mathrm{Res}(yx^{-1}dx) = \mathrm{tr}_{W/x(W)}(\pi - x\pi x^{-1})y.$$

Since $\pi - x\pi x^{-1}$ is the identity on $W/x(W)$, (1.4.13) follows. □

In our application of the above results, we will always have $x(W) \subseteq W$, so $(*)$ in principle gives a finite calculation for the residue form. Most of the time, we can actually use (1.4.13).

Our final results relate to extensions of the algebra K. The main theorem is

Theorem 1.4.14. *Let K be a k-algebra, V a K-module, and $W \subseteq V$ a near submodule. Suppose that $K \subseteq K'$, where K' is a commutative k-algebra that has a K-basis $\{x_1, \ldots, x_n\}$. Put*

$$V' := K' \otimes_K V \quad \text{and} \quad W' := \sum_{i=1}^{n} x_i \otimes W.$$

Then W' is a near K'-submodule of V' whose $\sim$-equivalence class is independent of the choice of K-basis for K', and for $y \in K'$ and $x \in K$ we have

$$\mathrm{Res}_{W'}^{V'}(y\,dx) = \mathrm{Res}_W^V(\mathrm{tr}_{K'/K}(y)dx).$$

Proof. If we put

$$\tilde{W} := \sum_{j=1}^{n} a_j x_j \otimes W - \sum_{j=1}^{n} x_j \otimes a_j W,$$

for any $a_j \in K$ $(1 \leq j \leq n)$, we have $\tilde{W} \preceq W$ by (1.4.6). From this it follows easily that W' is a near K'-submodule whose $\sim$ equivalence class is well-defined.

Now choose a projection $\pi : V \to W$. Since

$$V' = \bigoplus_{i=1}^{n} x_i \otimes V \quad \text{and} \quad W' = \bigoplus_{i=1}^{n} x_i \otimes W,$$

we can let $\pi_i := 1 \otimes \pi : x_i \otimes V \to W$ and define the projection $\pi' := \sum_i \pi_i : V' \to W'$.

Let $w = \sum_i x_i \otimes w_i \in W'$. Since $x \in K$, we have

$$xw = \sum_i x_i \otimes xw_i \quad \text{and} \quad \pi' xw = \sum_i x \otimes \pi x w_i.$$

For $y \in K'$, there exist $y_{ij} \in K$ with

$$yx_i = \sum_j y_{ij} x_j,$$

whence

$$yw = \sum_{i,j} y_{ij} x_j \otimes w_i = \sum_j x_j \otimes \sum_i y_{ij} w_i.$$

It follows that

$$[\pi' x, y](x_i \otimes w_i) = \sum_j x_j \otimes [\pi x, y_{ij}] w_i.$$

In particular, put $U_j := \Sigma_i [\pi x, y_{ij}](W)$ and

$$U := \bigoplus_j x_j \otimes U_j.$$

Then $[\pi' x, y](W') \subseteq U$, so U is a core subspace for $[\pi' x, y]$ by (1.4.9), and we conclude that

$$\begin{aligned}
\operatorname{tr}_{V'} [\pi' x, y] &= \operatorname{tr}_U \sum_{i,j} 1 \otimes [\pi x, y_{ij}] \\
&= \sum_j \operatorname{tr}_{U_j} 1 \otimes [\pi x, y_{jj}] \\
&= \operatorname{tr}_V [\pi x, \operatorname{tr}_{K'/K}(y)].
\end{aligned}$$

Since the residue form is antisymmetric, the result follows. □

Some care needs to be taken when extending K, because all our results have assumed a fixed ground field k. Suppose, however, that k' is a finite extension of k, and in (1.4.14) we have $K' = k' \otimes_k K$. Then V' and W' are actually k'-spaces, and we are often interested in computing traces with respect to k' rather than k. If x is any finitepotent operator on the k-vector space V, it remains finitepotent on $V' := k' \otimes_k V$, and just as in the finite-dimensional case, its k'-trace on V' is the same as its k-trace on V. Thus, we have

Lemma 1.4.15. *Suppose that W is a near submodule of the K-module V and that k' is a finite extension of k. Put $K' := k' \otimes_k K$ with k' and K identified with their natural images in K'. Then $W' := k' \otimes_k W$ is a near K-submodule of $V' := k' \otimes_k V$, and for $x, y \in K$ we have*

$$\langle y, x \rangle'_{V', W'} = \langle y, x \rangle_{V, W},$$

where the residue form $\langle x, y \rangle'$ is computed by taking k'-traces. □

It may happen that K already contains a copy of k', and that W is k'-invariant. Here, the k-trace on V and the k'-trace on V are related via the field trace $\operatorname{tr}_{k'/k}$.

Lemma 1.4.16. *Suppose that K contains a finite extension k' of k, and that the near K-submodule W of V is k'-invariant. Then*

$$\langle y,x\rangle_{V,W} = \mathrm{tr}_{k'/k}(\langle y,x\rangle'_{V,W}),$$

where the residue form $\langle x,y\rangle'$ is computed by taking k'-traces.

Proof. Since V is a K-module, it is a k'-vector space, and we are assuming that W is k'-invariant. Since the residue form is independent of the choice of projection map π, we can compute $\langle y,x\rangle_W$ using a k'-linear projection π. Since y and x commute with k', the map $[\pi y,x]$ is k'-linear. Now if U is any finite-dimensional k'-vector space and $f : U \to U$ is k'-linear, then by restriction f is also k-linear and we have $\mathrm{tr}_k(f) = \mathrm{tr}_{k'/k}(tr_{k'}(f))$. The formula follows. □

1.5 Exercises

Exercise 1.1. Let G be any totally ordered group, and let g_X, g_Y be any two elements of G. Define a map $v : k[X,Y]^\times \to G$ via

$$v(\sum_{i,j} a_{ij}X^iY^j) = \min_{a_{ij}\neq 0}(ig_X + jg_Y).$$

(i) Show that v is multiplicative and satisfies the ultrametric inequality (1.1.1). Conclude that v extends to a valuation on $k(X,Y)$ via $v(f/g) = v(f) - v(g)$.

(ii) Take $g_X = g_Y = 1 \in \mathbb{Z}$. Show that v is discrete. What is F_v?

(iii) Totally order $\mathbb{Z}\oplus\mathbb{Z}$ lexicographically, and put $g_X = (1,0)$ and $g_Y = (0,1)$. Show that $P_v = (Y)$ and $F_v = k$, but $Q := (XY^{-i} \mid i \geq 0)$ is a prime ideal that is not finitely generated.

(iv) Take $g_X = 1$, $g_Y = \sqrt{2} \in \mathbb{R}$. Show that $F_v = k$ and that P_v is the unique prime ideal of $\mathcal{O}_v$ but that P_v is not finitely generated.

Exercise 1.2. Define $\tau \in k((t))$ via

$$\tau = \sum_{i=0}^{\infty} t^{i!}.$$

Prove that the map $f(x,y) \mapsto f(t,\tau)$ defines an embedding $k(x,y) \to k((t))$. Thus, there is a discrete valuation on $k(x,y)$ with residue field k. Show that this valuation is not obtained by the construction of Exercise 1.1. [Hint: τ and t are algebraically independent because τ can be very well approximated by a polynomial of arbitrarily high degree.]

Exercise 1.3. Let $\mathcal{V}$ be a finite set of discrete valuations of a field K. Show that the field of fractions of $K(\mathcal{V};0)$ is K.

Exercise 1.4. Let $K = \mathbb{Q}(x,y)$ where $x = y^2(1-x^2+x^3)$. This is the example of the text after a change of variable. Let v be the valuation on $\mathbb{Q}(x)$ at $x = 0$. Show that v is ramified in K and that y is a local parameter at the unique divisor v' of v. Expand x in powers of y through y^6.

Exercise 1.5. Show that if v_P is the valuation of K afforded by $\mathscr{O}_P$ and if, for all $x \in K$, we define

$$|x|_v = b^{-v_P(x)}$$

for any fixed real number $b > 1$, then K becomes a normed field and $\hat{K}_P$ is just its metric space completion.

Exercise 1.6. Suppose that $\mathscr{O}_P$ is a discrete valuation ring of K. Show that $\hat{K}_P = K + \hat{\mathscr{O}}_P$.

Exercise 1.7. Suppose R is complete at I and R/I is a ring direct sum $R/I = S_1 \oplus S_2$. Show that R is a ring direct sum $R = R_1 \oplus R_2$ with $R_i/(R_i \cap I) = S_i$ $(i = 1,2)$. [Hint: This result is sometimes referred to as "lifting idempotents."]

Exercise 1.8. Suppose that $\mathscr{O}$ is a complete discrete valuation ring with maximal ideal P and field of fractions K, and that K' is a finite extension of K.

(i) Let R be the integral closure of $\mathscr{O}$ in K'. Generalize the argument of (1.2.11) to show that R is a complete free $\mathscr{O}$-module of finite rank.

(ii) Use Exercise 1.7 and (1.1.16) to deduce that there is a unique extension $(\mathscr{O}',P')$ of $(\mathscr{O},P)$ to K'.

(iii) Conclude that K' is complete at P', and that $|K' : K| = e(P'|P)f(P'|P)$.

Exercise 1.9. Let $R := k[[t]]$ and let $R_n := R/(t^n)$ for all $n \geq 0$. Define

$$\tilde{\Omega}_R := \overleftarrow{\lim}_n \Omega_{R_n/k}.$$

Prove the following:

(i) There is a natural derivation $d : R \to \tilde{\Omega}_R$ that is universal with respect to continuous k-derivations of R into complete R-modules.

(ii) $\tilde{\Omega}_R$ is a free R-module of rank 1 with basis dt.

Exercise 1.10. Let K be a k-algebra, and let I be the kernel of the map $K \otimes_k K \to K$ induced by multiplication. Show that the map $d : K \to I/I^2$ defined by

$$d(x) = x \otimes 1 - 1 \otimes x \qquad \text{mod } I^2$$

is a derivation and that the induced map $\Omega_{K/k} \to I/I^2$ is an isomorphism.

Exercise 1.11. Let K be a k-algebra and let X be an indeterminate. Show that $\Omega_{K[X]/k} = \Omega_{K/k} \oplus K dX$. If f is a polynomial in n variables over k, obtain the formula

$$df = \sum_{i=1}^{n} \frac{\partial f}{\partial X_i} dX_i.$$

Exercise 1.12. Prove formula (1.3.10).

Exercise 1.13. Let K be a field of characteristic $p > 0$ and let q be a power of p. Let $x \in K$ be a separating variable. For any $y \in K$, prove that

$$D_x^{(i)}(y^q) = \begin{cases} (D_x^{(i/q)}(y))^q & \text{if } i \equiv 0 \mod q, \\ 0 & \text{otherwise.} \end{cases}$$

Exercise 1.14. Prove that a linear operator is finitepotent if and only if it is the sum of a nilpotent operator and an operator of finite rank.

Exercise 1.15. (G.M. Bergman) In this exercise we will construct two trace zero operators whose sum has trace one.

Let W be a k-vector space with a countable basis $= \{e_0, e_1, \dots\}$. Let $R(e_i) = e_{i+1}$ and $L(e_i) = e_{i-1}$, $L(e_0) = 0$ be the right and left shift operators, respectively.

(i) Show that $LR = I$ and $RL = I - \pi$, where π is the natural projection onto e_0.

(ii) Let $V = W \oplus W \oplus W$ and define linear operators

$$P := \begin{bmatrix} 0 & R-I & 0 \\ 0 & 0 & R-I \\ 0 & 0 & 0 \end{bmatrix}, \quad Q := \begin{bmatrix} 0 & 0 & 0 \\ I & I & I \\ -L & -I & -I \end{bmatrix}$$

on V. Show that $P^3 = Q^3 = 0$.

(iii) Verify the following:

$$\begin{bmatrix} I & 0 & R \\ 0 & I & 0 \\ 0 & 0 & I \end{bmatrix} \begin{bmatrix} 0 & R-I & 0 \\ I & I & R \\ -L & -I & -I \end{bmatrix} \begin{bmatrix} I & 0 & -R \\ 0 & I & 0 \\ 0 & 0 & I \end{bmatrix} = \begin{bmatrix} -I+\pi & -I & 0 \\ I & I & 0 \\ -L & -I & 0 \end{bmatrix}.$$

(iv) Show that the right-hand side of the above is the sum of a nilpotent matrix and a rank 1 projection. Conclude that $P+Q$ is finitepotent of trace one.

2

Function Fields

In this chapter we make the basic assumption that K is a finitely generated extension of k of transcendence degree one. If $x \in K$ is any transcendental element, then $K/k(x)$ will be a finitely generated algebraic extension, i.e., a finite extension. Furthermore, we assume that k is algebraically closed in K, that is, that every element of K algebraic over k already lies in k. In this situation, we say that K is a *function field* over k, or sometimes that K/k is a function field.

2.1 Divisors and Adeles

By a *prime divisor* of K we shall mean the maximal ideal P of some k-valuation ring of K. We denote the corresponding valuation[1] by v_P and the residue field by F_P. By (1.1.14) and (1.1.19), all k-valuations of K are discrete. This is a critical fact upon which the entire subsequent development depends. We let $\mathbb{P}_K$ be the set of all prime divisors of K.

Let $x \in K$ and suppose that $v_P(x) \geq 0$ for some prime divisor P. Then we say that "x is finite at P" and define $x(P) \in F_P$ to be the residue $x + P \mod P$. Thus x vanishes at P iff $v_P(x) > 0$, in which case we say that "x has a zero of order $v_P(x)$ at P." If x is not finite at P, then we say that "x has a pole of order $-v_P(x)$ at P."

Lemma 2.1.1. *Let P be a prime divisor of K and suppose that $x \in K$ vanishes at P. Then v_P divides the x-adic valuation v_x of $k(x)$. In particular, F_P is a finite extension of k of degree $f(v_P|v_{(x)})$.*

---

[1] Some authors use the notation ord_P here.

Proof. Since $\nu_P(x) > 0$, it follows immediately from (1.1.14) that $\nu_P \mid \nu_x$. Since the residue field of ν_x is just k, the result follows. □

We write $\deg(P) := |F_P : k|$ for the *degree* of P. Note that the residue degree of ν_P over ν_x is independent of x, and if k is algebraically closed, all prime divisors have degree one. Some care needs to be taken when evaluating a function x at a prime P of degree greater than one. The reason is that there is no natural embedding of F_P into any given algebraic closure of the ground field. So, for example, the question of whether $x(P) = x(Q)$ is not really well-defined in general unless P and Q have degree one.

We will refer to prime divisors of degree one as "points" because in the algebraically closed case they correspond to points of the unique nonsingular projective curve whose function field is K. We will study this case in detail in Chapter 4. When k is not algebraically closed, the question of whether K has any points is interesting.

Lemma 2.1.2. *If P_i is a prime divisor of K of degree f_i and $x \in K^\times$ with $\nu_{P_i}(x) = e_i$ for $1 \le i \le s$, then*

$$\sum_{i=1}^{s} e_i f_i \le |K : k(x)|. \tag{2.1.3}$$

In particular, x has only finitely many zeros and poles.

Proof. If $x \in k$, both sides of the inequality are zero. For $x \notin k$, this is a straightforward application of (1.1.22), viewing K as a finite extension of $k(x)$. Namely, put $\nu_i := \nu_{P_i}$ and let $\nu_{(x)}$ be the valuation of $k(x)$ whose valuation ring is $k[x]_{(x)}$. Then the ramification index of ν_i over $\nu_{(x)}$ is precisely the order of the zero of x at P_i, and the degree of P_i is precisely the residue degree of ν_i over $\nu_{(x)}$. □

One of the important results in this section is to show that the above inequality is actually an equality when all zeros and poles of x are included, but first we need some machinery.

A *divisor* on K is an element of the free abelian group generated by the prime divisors, that is, it is a formal finite integral linear combination of prime divisors. We denote this group by $\mathrm{Div}(K)$. We define the *degree* of a divisor $D := \sum_P d_P P$ to be $\deg(D) := \sum_P d_P \deg(P)$ and we define $\nu_P(D) := d_P$ for any valuation ν_P. Thus, $\nu_P(D) = 0$ for almost all P.

For an element $x \in K^\times$, the *principal divisor* of x is the divisor $[x] := \sum_P \nu_P(x)P$. Note that the sum is finite by (2.1.2). It is often convenient to distinguish the positive and negative terms in this sum. So we define the *zero divisor* (resp. *pole divisor*) of any divisor $D := \sum_P d_P P$ to be $D_0 := \sum_{d_P > 0} d_P P$ (resp. $D_\infty := D_0 - D$).

An important property of divisors is that a function in K is uniquely determined by its principal divisor, up to a constant multiple.

Lemma 2.1.4. *Any nonconstant element of K has at least one zero and one pole. Hence, any two elements of K with the same divisor differ by a constant multiple.*

Proof. Since any nonconstant function in K is transcendental over k, (1.1.7) yields prime divisors P, Q with $v_P(x) > 0$ and $v_Q(x^{-1}) > 0$, so x has a zero at P and a pole at Q. Since $[xy] = [x] + [y]$, we see that $[x] = [y]$ implies that $xy^{-1} \in k$. □

Since $v(xy) = v(x) + v(y)$, the principal divisors form a subgroup of the group of divisors. The quotient group is called the *divisor class group*. We say that two divisors are *linearly equivalent* and write $D \sim D'$ if $D - D' = [x]$ for some principal divisor $[x]$.

The divisors are partially ordered by setting $D \leq D'$, provided that $v(D) \leq v(D')$ for all valuations v. A divisor D with $D \geq 0$ is called nonnegative, or sometimes *effective*. We can now make the following fundamental

Definition. Let K be a function field and D a divisor on K. Then

$$L_K(D) := \{x \in K^\times \mid [x] \geq -D\} \cup \{0\}.$$

We will write $L(D)$ for $L_K(D)$ when there is no danger of confusion. The following properties of $L(D)$ are straightforward:

Lemma 2.1.5. *Let D be a divisor on K. Then*

1. *$L(D)$ is a k-linear subspace of K.*
2. *If $D_1 \sim D_2$, then $L(D_1) \cong L(D_2)$.*
3. *$L(D) \neq 0$ iff there is a nonnegative divisor $D' \sim D$.*

Proof. 1. This follows from the ultrametric inequality (1.1.1).

2. Suppose $D_2 = D_1 + [x]$, then multiplication by x is an isomorphism from $L(D_1)$ to $L(D_2)$.
3. $[x] \geq -D$ iff $[x] + D \geq 0$. □

An important fact, which we shall prove shortly, is that $L(D)$ is finite-dimensional. Note that if $D_0 = \sum_P a_P P$ and $D_\infty = \sum_P b_P P$, then the condition $x \in L(D)$ is equivalent to the following two conditions:

1. x can have a pole at a prime P only if $a_P > 0$, and the order of that pole can be at most a_P,
2. x has a zero of order at least b_P at P for all P.

Note that $k \subseteq L(D)$ iff $D \geq 0$.

It may be instructive to illustrate the preceding ideas in the case $K = k(X)$ before proceeding. Let $\mathbb{P} := \mathbb{P}_{k(X)}$, $P_\infty := (1/X)$, and let $\mathbb{P}_0 := \mathbb{P} \setminus \{P_\infty\}$. The set $\mathbb{P}_0$ is in one-to-one correspondence with the set of monic irreducible polynomials $P(X) \in k[X]$. Recall that $\mathcal{O}_P$ is just the localization of $k[X]$ at the prime ideal generated by $P(X)$. It is easy to verify that in this case $F_P \simeq k[X]/(P(X))$ and therefore the degree of the prime divisor P is just the degree of the polynomial $P(X)$.

Let $f(X) := \prod_{i=1}^{r} P_i(X)^{e_i} \in k(X)$. Then $v_\infty(f) = -\deg f = -\sum_i e_i \deg P_i$, and the principal divisor of f is

$$[f] = \sum_i e_i P_i - \Big(\sum_i e_i \deg P_i\Big) P_\infty.$$

So we see that $\deg[f] = 0$, and also that conversely, every divisor of degree zero is principal. The first property generalizes to all function fields, while the second turns out to be characteristic of $k(X)$. What is the subspace $L(nP_\infty)$? It consists of all rational functions having a pole only at infinity, of order at most n. This is just the set of polynomials of degree at most n (see Exercise 2.2). We see that $\dim L(nP_\infty) = \deg nP_\infty + 1$. By Exercise 2.1, this statement remains true for $k(X)$ when nP_∞ is replaced by any nonnegative divisor. The generalization of this statement to an arbitrary function field is Riemann's theorem: "1" must be replaced by some other integer depending only on K, and then the equality holds for all divisors of sufficiently large degree.

For $P \in \mathbb{P}_K$, let $\hat{O}_P$ denote the completion of the local integers $\mathscr{O}_P$ at P (see Section 1.2). We denote the field of fractions of $\hat{\mathscr{O}}_P$ by $\hat{K}_P$. By (1.2.10) the residue fields of $\mathscr{O}_P$ and $\hat{\mathscr{O}}_P$ are canonically isomorphic. We denote them by F_P.

We next define the *adele ring* of K, A_K, to be the subring of the direct product $\prod_{P \in \mathbb{P}_K} \hat{K}_P$ consisting of all tuples $\{\alpha_P \mid P \in \mathbb{P}_K\}$ such that $v_P(\alpha_P) \geq 0$ for almost all P. Addition and multiplication in A_K are defined component-wise.

We identify K with its natural image in the direct product, and extend valuations v_P to A_K by defining $v_P(\alpha) := v_P(\alpha_P)$. This allows us to define, for any divisor D,

$$A_K(D) := \{\alpha \in A_K \mid v(\alpha) \geq -v(D) \quad \text{for all } v\}.$$

Again, the ultrametric inequality (1.1.1) shows that $A_K(D)$ is a k-linear subspace of A_K. Moreover, $A_K(D) \cap K = L(D)$.

Lemma 2.1.6. *Suppose $D_1 \leq D_2$ are divisors on K. Then $A_K(D_1) \subseteq A_K(D_2)$, and*

$$\dim(A_K(D_2)/A_K(D_1)) = \deg(D_2) - \deg(D_1).$$

Proof. It is immediate from the definitions that $A_K(D_1) \subseteq A_K(D_2)$. By induction on $\deg(D_2) - \deg(D_1)$, we may assume that $D_2 = D_1 + P$ for some prime divisor P, and prove that $\dim(A_K(D_2)/A_K(D_1)) = \deg P$.

Let t be a local parameter at P and let F_P be the residue field. Put $e := v_P(D_2)$, and consider the k-linear mapping $\phi : A_K(D_2) \to F_P$ given by $\phi(\alpha) := t^e \alpha_P + P$. It is immediate that $\ker(\phi) = A_K(D_1)$. On the other hand, for any element $x + P \in F_P$, there is an adele α with $\alpha_P = xt^{-e}$ and $\alpha_{P'} = 0$ for $P' \neq P$, whence $\phi(\alpha) = x + P$. Thus, ϕ induces a k-isomorphism $A_K(D_2)/A_K(D_1) \cong F_P$. □

Given two divisors D_1 and D_2 we let $D_1 \cup D_2$ (resp. $D_1 \cap D_2$) denote their least upper bound (resp. greatest lower bound) with respect to the partial order $\leq$. In other words, $v(D_1 \cup D_2) := \max\{v(D_1), v(D_2)\}$ and $v(D_1 \cap D_2) := \min\{v(D_1), v(D_2)\}$ for all valuations v.

Lemma 2.1.7. *Given any two divisors D_1, D_2 we have*

1. $A_K(D_1 \cap D_2) = A_K(D_1) \cap A_K(D_2)$,
2. $A_K(D_1 \cup D_2) = A_K(D_1) + A_K(D_2)$.

Proof. 1. This is immediate from the definitions: $\alpha \in A_K(D_1 \cap D_2)$ iff $-v(\alpha) \leq v(D_1 \cap D_2) = \min\{v(D_1), v(D_2)\}$ for all v iff $\alpha \in A_K(D_1) \cap A_K(D_2)$.

2. By (2.1.6) we have $A_K(D_1) + A_K(D_2) \subseteq A_K(D_1 \cup D_2)$. From the definitions we obtain

$$\deg D_1 - \deg(D_1 \cap D_2) = \deg(D_1 \cup D_2) - \deg D_2.$$

Now using (2.1.6) again, a dimension count yields

$$\begin{aligned}\dim(A_K(D_1) + A_K(D_2))/A_K(D_2) &= \dim A_K(D_1)/A_K(D_1 \cap D_2) \\ &= \deg D_1 - \deg D_1 \cap D_2 \\ &= \deg D_1 \cup D_2 - \deg D_2 \\ &= \dim A_K(D_1 \cup D_2)/A_K(D_2). \qquad \square\end{aligned}$$

The quotient space A_K/K is a k-vector space, which is called the *adele class group*. Given an adele α or a subspace $V \subseteq A_K$, we denote by $\overline{\alpha}$ or $\overline{V}$ its image in the adele class group.

Lemma 2.1.8. *Suppose $D_1 \leq D_2$ are divisors on K. Then there is a natural short exact sequence*

$$0 \to L(D_2)/L(D_1) \to A_K(D_2)/A_K(D_1) \to \overline{A_K(D_2)}/\overline{A_K(D_1)} \to 0. \tag{2.1.9}$$

Proof. This is an exercise in using the isomorphism theorems[2]. Let

$$\phi : A_K(D_2) \to \overline{A_K(D_2)}$$

be the natural map, with kernel $L(D_2)$. Then $\phi^{-1}(\overline{A_K(D_1)}) = L(D_2) + A_K(D_1)$. So the kernel of the map $A_K(D_2)/A_K(D_1) \to \overline{A_K(D_2)}/\overline{A_K(D_1)}$ induced by ϕ is

$$(L(D_2) + A_K(D_1))/A_K(D_1) \simeq L(D_2)/(L(D_2) \cap A_K(D_1)) = L(D_2)/L(D_1). \qquad \square$$

Corollary 2.1.10. *$L(D)$ is finite dimensional, for any divisor D. If $D_1 \leq D_2$ are divisors, then*

$$\dim L(D_2)/L(D_1) \leq \deg D_2 - \deg D_1. \tag{2.1.11}$$

In particular, if D is any nonnegative divisor, then

$$\dim L(D) \leq \deg D + 1. \tag{2.1.12}$$

---

[2] It is also immediate from the "nine-lemma" of homological algebra.

Proof. Inequality (2.1.11) is immediate from (2.1.6) and (2.1.9). Setting $D_2 = D$ and $D_1 = 0$, we have $\dim L(D) - \dim L(0) \leq \deg L(D)$. But $L(0) = k$ by (2.1.4), whence $L(D)$ has finite dimension at most equal to $\deg D + 1$. □

These inequalities are quite important. We next investigate the extent to which (2.1.12) fails to be an equality. To this end, we define $\delta(D) = \deg D + 1 - \dim L(D)$ for any divisor D. The following important property is immediate from the short exact sequence:

Corollary 2.1.13. *Let $D_1 \leq D_2$ be divisors on K. Then*

$$\delta(D_2) - \delta(D_1) = \dim(\overline{A_K(D_2)}/\overline{A_K(D_1)}).$$

In particular, $\delta(D_1) \leq \delta(D_2)$. □

The main point of this section is to prove that $\delta(D)$ is a constant for all divisors D of sufficiently large degree. In particular, this will show that $L(D) \neq 0$ for all such D. As a first step in that argument, we show that $\delta([x^m]_\infty)$ is bounded as a function of m for all $x \in K$. This result has several important consequences, among them the fact that principal divisors have degree zero. This result is sometimes called the *product formula* for function fields.

Theorem 2.1.14. *Let $x \in K$. Then* $\deg[x] = 0$, *and there is an integer B depending on x such that $\delta([x^m]_\infty) \leq B$ for all $m \geq 0$. Furthermore, if $[x]_0 = \sum_{i=1}^r e_i P_i$, then*

$$\deg[x]_\infty = \deg[x]_0 = \sum_{i=1}^r e_i \deg P_i = |K : k(x)|. \tag{2.1.15}$$

Proof. Let $\{u_1, \dots, u_n\}$ be a basis for $K/k(x)$, and let D be a nonnegative divisor such that $[u_j] \geq -D$ for all j. Thus $u_j \in L(D)$ for all j. For any positive integer m, the functions $u_i x^j$ $(1 \leq i \leq n,\ 0 \leq j < m)$ are linearly independent over k and lie in $L([x^m]_\infty + D)$. By (2.1.12) we have

$$mn \leq \dim L([x^m]_\infty + D) \leq m \deg[x]_\infty + \deg(D) + 1 \tag{$*$}$$

for all m. It follows that $\deg[x]_\infty \geq n = |K : k(x)|$ for all nonconstant $x \in K$. Since $k(x) = k(x^{-1})$, we also have $\deg[x]_0 \geq |K : k(x)|$. Now (2.1.3) implies that

$$\deg[x]_0 = \deg[x]_\infty = |K : k(x)|,$$

whence (2.1.15). In particular, it follows that $\deg[x] = 0$. Finally, since $\deg[x^m]_\infty = mn$, we can use (2.1.11) and $(*)$ to obtain

$$\begin{aligned}\delta([x^m]_\infty) &= 1 + mn - \dim L([x^m]_\infty) \leq 1 + \dim L([x^m]_\infty + D) - \dim L([x^m]_\infty) \\ &\leq 1 + \deg([x^m]_\infty + D) - \deg[x^m]_\infty = 1 + \deg D.\end{aligned}$$ □

The fact that principal divisors have degree zero is fundamental. Note that this is immediate for the rational function field $k(X)$ by (1.1.14). There are some important corollaries, the first of which is straightforward.

Corollary 2.1.16.

1. *If* $D \sim D'$, *then* $\deg D = \deg D'$ *and* $\delta(D) = \delta(D')$.
2. $L(D) = 0$ *for all divisors* D *with* $\deg D < 0$.

Proof.

1. If $D' = D + [x]$ for some principal divisor $[x]$, then $\deg D' = \deg D + \deg[x] = \deg D$, and $\dim L(D) = \dim L(D')$ by (2.1.5).
2. If $0 \neq x \in L(D)$, then $D + [x] \geq 0$ and in particular, $\deg D = \deg(D + [x]) \geq 0$. □

More importantly, we can now show that the inequality of (1.1.22) is an equality.

Corollary 2.1.17. *Let* K/k *be a function field and let* K' *be a finite extension of* K. *Suppose that* P *is a prime divisor of* K, *and let* $Q_1, \dots, Q_r$ *be the set of all distinct primes of* K' *dividing* P. *Then*

$$\sum_{i=1}^{r} e(Q_i|P) f(Q_i|P) = |K' : K|.$$

Proof. Choose $0 \neq x \in P$, let $\{P = P_1, P_2, \dots, P_s\}$ be the set of all prime divisors of (x) in K and let $e_i := e(P_i|(x))$. The P_i are the zeros of x in $\mathbb{P}_K$.

Let $P_{ij} (1 \leq j \leq r_i)$ be all the prime divisors of P_i in K', so $r_1 = r$ and $Q_j = P_{1j}$ for $1 \leq j \leq r$. Put $e_{ij} := e(P_{ij}|P_i)$ for all i, j. Then the P_{ij} are the zeros of x in $\mathbb{P}_{K'}$. By (1.1.25) applied to the tower $k(x) \subseteq K \subseteq K'$, we have $e(P_{ij}|(x)) = e_{ij} e_i$.

Let F_i (resp. F_{ij}) be the residue field at P_i (resp. P_{ij}). Then $\deg P_i = |F_i : k|$. However, an important point to keep in mind is that K' may contain additional elements algebraic over k. Let k' be the set of all such elements. Then K' is a function field over k', and $\deg P_{ij} = |F_{ij} : k'|$. By (1.1.22) we have

$$n_i := \sum_j e_{ij} |F_{ij} : F_i| \leq |K' : K|$$

for all i. We are trying to prove that the inequality is an equality for $i = 1$. Using (2.1.14) we get

$$\begin{aligned}
|K : k(x))| &= \sum_i e_i |F_i : k|, \quad \text{and} \\
|K' : k(x)| &= |k' : k| |K' : k'(x)| = |k' : k| \sum_{ij} e_{ij} e_i |F_{ij} : k'| \\
&= \sum_{ij} e_{ij} e_i |F_{ij} : F_i| |F_i : k| = \sum_i n_i e_i |F_i : k| \leq |K' : K| |K : k(x)|.
\end{aligned}$$

It follows that $n_i = |K' : K|$ for all i. □

As a further consequence, we obtain

Corollary 2.1.18. *Let P be a prime of K. Then the integral closure of $\mathscr{O}_P$ in K' is a finitely generated $\mathscr{O}_P$-module.*

Proof. This is immediate from (1.1.8), (2.1.17) and (1.1.22). □

We can now prove a preliminary version of the Riemann–Roch theorem.

Theorem 2.1.19 (Riemann). *There exist positive integers N and g depending only on K such that $\delta(D) \leq g$ for all divisors D, with equality holding for all divisors of degree at least N.*

Proof. Fix a nonconstant function $x \in K$. We first argue that for any divisor D, there is an equivalent divisor D' and a positive integer m such that $D' \leq [x^m]_\infty$. Namely, since $D_0 \geq 0$ we have $[x^m]_\infty - D_0 \leq [x^m]_\infty$ for all positive integers m. Then (2.1.11) implies that

$$\dim L([x^m]_\infty) - \dim L([x^m]_\infty - D_0) \leq \deg [x^m]_\infty - \deg([x^m]_\infty - D_0) = \deg D_0.$$

Since $\delta([x^m]_\infty)$ is bounded as a function of m, it follows that $\dim L([x^m]_\infty - D_0) > 0$ for sufficiently large m. For such an m, choose a nonzero element y of $L([x^m]_\infty - D_0)$. Then

$$[y] \geq D_0 - [x^m]_\infty \geq D - [x^m]_\infty,$$

whence $D' := D - [y] \leq D_0 - [y] \leq [x^m]_\infty$ as claimed.

Now using (2.1.16) and (2.1.13) we have

$$\delta(D) = \delta(D') \leq \delta([x^m]_\infty)$$

for a suitably large positive integer m. This shows that $\delta(D)$ is bounded, for all divisors D.

Let $g := \operatorname{lub}\{\delta(D) \mid D \text{ any divisor }\}$, and choose a divisor D' with $\delta(D') = g$. Put $N := \deg D' + g + 1$, and let D be any divisor of degree at least N. Then $\deg(D - D') > g$, but $\delta(D - D') \leq g$. This implies that $L(D - D') \neq 0$. Taking $x \in L(D - D')$ we have $[x] \geq D' - D$ whence

$$g \geq \delta(D) = \delta(D + [x]) \geq \delta(D') = g. \qquad \square$$

The integer $g = g_K$ above is called the *genus* of the function field K. Perhaps the main point of Riemann's theorem is that it guarantees a nonconstant function in $L(D)$ for any divisor D of suitably large degree. In fact for degree at least $g+1$ we have $\dim L(D) \geq \deg D - g + 1 \geq 2$. So for example, if P is a prime divisor there is a function $x \in L((g+1)P)$ with exactly one pole, namely at P, and that pole has multiplicity at most $g+1$.

2.2 Weil Differentials

Here we refine Riemann's theorem by looking more closely at divisors D for which $\delta(D) < g$. Such divisors are called *special*.

Theorem 2.2.1. *If K has genus g, then*

$$\dim A_K/(A_K(D)+K) = g - \delta(D)$$

for any divisor D of K.

Proof. Let D be a nonspecial divisor and let α be an adele. There is certainly a divisor $D_1 \geq D$ such that $\alpha \in A_K(D_1)$. From (2.1.13) we see that D_1 is also nonspecial and that $A_K(D)+K = A_K(D_1)+K$. Thus,

$$\alpha \in A_K(D_1) \subseteq A_K(D_1)+K = A_K(D)+K,$$

which shows that $A_K(D)+K = A_K$ for D nonspecial. But given any divisor D we can choose $D_1 \geq D$ of sufficiently large degree so that D_1 is nonspecial. Then $A_K(D_1)+K = A_K$ and $\dim(A_K(D_1)+K)/(A_K(D)+K) = \delta(D_1) - \delta(D)$ by (2.1.13). □

We call $g - \delta(D)$ the *index of speciality* of D.

Further mileage may be obtained by looking at the dual of A_K. We define a *Weil differential* on K to be a k-linear functional on A_K that vanishes on $A_K(D)+K$ for some divisor D. The use of the word "differential" will be justified later. Denote by W_K the space of Weil differentials, and let $W_K(D)$ be the subspace of those differentials which vanish at $A_K(D)+K$. Note that if $D_1 \leq D_2$, then $W_K(D_1) \supseteq W_K(D_2)$. Moreover, by (2.2.1) we have

(2.2.2) $$\dim W_K(D) = g - \delta(D).$$

In particular, (2.1.19) implies that $W_K(D) = 0$ for all D of sufficiently large degree. Then if we fix a nonzero $w \in W_K$, we can choose a divisor D of maximal degree such that $w \in W_K(D)$.

Lemma 2.2.3. *Let w be a nonzero Weil differential. Then there is a unique divisor D of maximum degree such that $w \in W_K(D)$. Moreover, for any divisor E we have $w \in W_K(E)$ iff $E \leq D$.*

Proof. This is an easy consequence of (2.1.7): If w vanishes on $A_K(D_1)$ and $A_K(D_2)$, then it vanishes on $A_K(D_1)+A_K(D_2) = A_K(D_1 \cup D_2)$. □

We define the divisor of a Weil differential w to be the unique divisor given by (2.2.3), and denote it by $[w]$. We define $v_P(w) := v_P([w])$. Let $P \in \mathbb{P}_K$ with local parameter t, and identify $\hat{K}_P$ with the set of all adeles α with $\alpha_Q = 0$ for $Q \neq P$. We observe that if e is an integer and D is any divisor with $v_P(D) \geq e$, then $t^{-e}\hat{\mathcal{O}}_P \subseteq A_K(D)$. Then directly from the definitions we have

Lemma 2.2.4. *For any prime divisor P and any integer e, we have $v_P(w) \geq e$ if and only if w vanishes on $t^{-e}\hat{\mathcal{O}}_P$. In particular, w restricts to a nonzero k-linear functional on $\hat{K}_P$.* □

As we will see, the restriction of w to $\hat{K}_P$ turns out to be the local residue map.

Any divisor of the form $[w]$ for some Weil differential w is called a *canonical divisor*. The interesting fact is that all canonical divisors are linearly equivalent. To prove this, we first observe that there is an action of K on W_K, given by $xw(\alpha) := w(x\alpha)$ for $x \in K$, $w \in W_K$, and $\alpha \in A_K$.

Lemma 2.2.5. *$L(D)W_K(C) \subseteq W_K(C-D)$ for any divisors C,D. Moreover, we have*

$$[xw] = [x] + [w] \tag{2.2.6}$$

for any $x \in K$ and $w \in W_K$.

Proof. It is immediate from the definitions that $A_K(C)A_K(D) \subseteq A_K(C+D)$ for any divisors C,D. Thus, for $x \in L_K(D)$ we have $xA_K(C) \subseteq A_K(C+D)$. This implies that $xW_K(C) \subseteq W_K(C-D)$.

In particular, since $x \in L(-[x])$, we have $[xw] \geq [x] + [w]$ for any $x \in K$ and $w \in W_K$. Substituting $x^{-1}w$ for w in this inequality yields

$$[w] \geq [x] + [x^{-1}w] \geq [x] - [x] + [w] = [w],$$

whence

$$[x^{-1}w] = [w] - [x],$$

and (2.2.6) follows. □

For $w \in W_K$ and $P \in \mathbb{P}_K$, we define

$$v_P(w) := v_P([w]).$$

Then the following properties are immediate:

$$\begin{aligned} v_P(xw) &= v_P(x) + v_P(w), \\ v_P(w+w') &\leq v_P(w) + v_P(w'), \end{aligned} \tag{2.2.7}$$

for all $x \in K$ and $w, w' \in W_K$.

Theorem 2.2.8. *Let K be a function field. Then* $\dim_K(W_K) = 1$. *Any two canonical divisors are linearly equivalent.*

Proof. The second statement is immediate from the first and (2.2.6). Choose any two nonzero Weil differentials w_1, w_2. For $i = 1,2$ suppose that $w_i \in W_K(D_i)$. Then the map $x \mapsto xw_i$ defines for any divisor D an embedding $\phi_{i,D} : L(D) \to W_K(D_i - D)$ by (2.2.5). Now let D be a divisor of large degree, and consider the pair of embeddings $\phi_{i,D+D_i} : L(D+D_i) \to W_K(-D)$. Note that $L(-D) = 0$, so we have

$$\dim W_K(-D) = g - \delta(-D) = g - \deg(-D) - 1 = g + \deg D - 1.$$

For each embedded subspace, however, we have

$$\dim L(D+D_i) = \deg D + \deg D_i - g + 1 \quad \text{for } i = 1,2,$$

so the codimension of these subspaces in $W_K(-D)$ is $2g-2-\deg D_i$, which is independent of $\deg D$. Hence the two subspaces intersect for D of suitably large degree. This means that there exist elements $x_i \in K$ with $x_1 w_1 = x_2 w_2$, as required. □

We are at last ready to prove the main theorem of this chapter.

Theorem 2.2.9 (Riemann–Roch). *Let K be a function field of genus g and let C be a canonical divisor on K. Then for any divisor D, $L_K(C-D) \simeq W_K(D)$, and we have*

$$\dim L_K(D) = \deg D + 1 - g + \dim L_K(C-D).$$

Proof. A restatement of the formula is $\dim L(C-D) = g - \delta(D)$, so by (2.2.2) the formula follows from the k-isomorphism $L(C-D) \simeq W_K(D)$.

By (2.2.8) we can take $C = [w]$ for any nonzero Weil differential w. Then $w \in W_K(C)$, so the map $x \mapsto xw$ embeds $L(C-D)$ into $W_K(D)$ by (2.2.5). To show that this map is onto, let $w' \in W_K(D)$. Then $[w'] \geq D$ by definition, and $w' = xw$ for some $x \in K$ by (2.2.8). Then (2.2.6) yields $[x] = [w'] - [w] \geq D - C$, whence $x \in L(C-D)$, and we have $L(C-D) \simeq W_K(D)$, as required. □

The Riemann–Roch theorem has many important consequences, which we will be exploring in subsequent sections. For now, we list a few of the more obvious ones.

Corollary 2.2.10. *Let K be a function field of genus g and let C be a divisor on K. Then C is a canonical divisor if and only if $\dim L(C) = g$ and $\deg C = 2g-2$. In particular, all divisors of degee at least $2g-1$ are nonspecial.*

Proof. Suppose C is canonical and put $D = 0$ in (2.2.9). This yields $\dim C = g$. Now put $D = C$ and obtain $\deg C = 2g-2$. Conversely, assume $\dim L(C) = g$ and $\deg C = 2g-2$. Then $\delta(C) = g-1$, so $\dim \overline{A_K / A_K(C)} = 1$ by (2.2.1). This means that there exists a nonzero Weil differential $w \in W_K(C)$. Then $C \leq [w]$, but $\deg C = \deg[w]$, and therefore $C = [w]$.

Finally, if $\deg D \geq 2g-1$, then $\deg(C-D) < 0$ and hence $L(C-D) = 0$, so D is nonspecial by (2.2.9). □

Corollary 2.2.11. *The following conditions are equivalent for a function field K:*

1. *K has genus 0 and has a prime divisor P of degree one.*
2. *K has an element x with $\deg[x]_\infty = 1$.*
3. *$K = k(x)$ for some $x \in K$.*

Proof. $1 \implies 2:$ By (2.2.10) canonical divisors have degree -2, so the Riemann–Roch theorem gives $\dim L(P) = 2$. Let x be a nonconstant function in $L(P)$.

Then x has exactly one pole of order at most 1 at P. But x must have a pole, so we have $[x]_\infty = P$.

$2 \implies 3$: We have $\deg[x]_\infty = 1 = |K : k(x)|$ by (2.1.14).

$3 \implies 1$: Clearly, $k(x)$ has the point (x). We have previously observed that, for $K = k(x)$, $L(nP_\infty)$ is the space of polynomials of degree at most n, and therefore $\dim L(nP_\infty) = n+1 = \deg(nP_\infty)+1$ for all n. Thus (2.2.9) implies that $g = 0$. □

Recall from (2.1.16) that the degree map is well-defined on divisor classes. The degree zero subgroup of the divisor class group is called the *Jacobian* of K, denoted $J(K)$. If K has a point (prime divisor of degree one) P_0, then there is an obvious map $\psi(P) = \bar{P} - \bar{P}_0$ from the points of K to $J(K)$, where $\bar{P}$ denotes the image of P in the divisor class group.

Corollary 2.2.12. *If $g(K) > 0$, then ψ is injective.*

Proof. The condition $P - P' = [x]$ implies that $[x]_\infty = P'$, so $g_K = 0$ by (2.2.11). □

The Riemann–Roch theorem yields the following improvement of the weak approximation theorem:

Theorem 2.2.13 (Strong Approximation Theorem). *Suppose that*

$$S := \{P_\infty, P_1, \dots, P_n\} \subseteq \mathbb{P}_K,$$

$\{x_1, \dots, x_n\} \subseteq K$, and $\{m_1, \dots, m_n\} \subseteq \mathbb{Z}$. Put $v_i := v_{P_i}$ for all i. Then there exists $x \in K$ such that $v_i(x - x_i) = m_i$ $(1 \le i \le n)$ and $v_P(x) \ge 0$ for all primes $P \notin S$.

Proof. Consider the divisor $D := NP_\infty - \sum_{i=1}^n (m_i + 1)P_i$ where $N \gg 0$, and the adele

$$\alpha_P := \begin{cases} x_i & \text{if } P = P_i,\ 1 \le i \le n, \\ 0 & \text{otherwise.} \end{cases}$$

For N sufficiently large, D is nonspecial, so $A_K = A_K(D) + K$ by (2.2.1). In particular, there is an element $y \in K$ with $y - \alpha \in A_K(D)$. This means that $v_i(y - x_i) \ge m_i + 1$ for all i, and $v_P(y) \ge 0$ for $P \notin S$.

Next, choose $z_i \in K$ with $v_i(z_i) = m_i$ for all i. Then repeating the above argument, there is an element $z \in K$ with $v_i(z - z_i) > m_i$ for all i, and $v_P(z) \ge 0$ for $P \notin S$. Then $v_i(z) = v_i(z - z_i + z_i) = m_i$ for all i. The element $x := y + z$ satisfies the conditions of the theorem. □

2.3 Elliptic Functions

The Riemann–Roch theorem has some very interesting consequences in the case that the genus of K is one. By (2.2.10), 0 is a canonical divisor, so the theorem now reads

$$\dim L(D) = \deg D + \dim L(-D).$$

For positive divisors D, this becomes $\dim L(D) = \deg D$. In addition to $g_K = 1$, suppose that K has a point P_0. Then $\dim L(nP_0) = n$ for all $n > 0$. Clearly, $L(P_0) = k$. Let $x \in L(2P_0) \setminus L(P_0)$ and let $y \in L(3P_0) \setminus L(2P_0)$. Then $[x]_\infty = 2P_0$ and $[y]_\infty = 3P_0$. By (2.1.14) we have $|K : k(x)| = 2$ and $|K : k(y)| = 3$. This implies that $K = k(x,y)$ and that y satisifies a quadratic polynomial over $k(x)$. In fact, the set $\{1, x, y, x^2, x^3, xy, y^2\} \subseteq L(6P_0)$ must be linearly dependent, because $\dim L(6P_0) = 6$. Moreover, the coefficient of y^2 in this dependence relation must be nonzero, or else we would have $K = k(x)$, and the coefficient of x^3 must be nonzero, or else $|K : k(y)| \leq 2$. Thus, we have proved

Theorem 2.3.1. *Let K be a function field of genus one with at least one point P_0. Then there is a basis $\{1, x, y\}$ for $L(3P_0)$ such that $K = k(x,y)$ and*

$$y^2 + f(x)y + g(x) = 0, \tag{2.3.2}$$

where $f(X) \in k[X]$ is linear and $g(X) \in k[X]$ is a cubic. □

Later, in (4.5.16), we will see, conversely, that any function field generated by elements x and y satisfying a cubic polynomial has genus at most one with equality if and only if a certain nonsingularity condition is satisfied.

Further simplifications can be made in the form of f and g, depending on whether or not $\operatorname{char}(k) = 2$, but we will not pursue this here.[3] A more interesting line of investigation starts from the observation that if D is any divisor of degree one, then there exists a nonzero $x \in L(D)$, so $D + [x]$ is nonnegative of degree one, i.e., a point. In other words, all divisors of degree one are linearly equivalent to points. Now let D be an arbitrary divisor of degree zero. Then $D + P_0 \sim P$ for some point P. Thus, we see that the map $\psi(P) = \bar{P} - \bar{P}_0$ of (2.2.12) is surjective, and therefore induces a bijective correspondence between the points of K and the Jacobian of K. We can then define a unique group operation $\oplus$ on the points of K making ψ an isomorphism. The zero element for this operation is just the point P_0. Note that the functions x and y are defined at all nonzero points P.

Recall that for any $P \in \mathbb{P}_K$ and any $f \in \mathcal{O}_P$, the image of f under the residue map $\mathcal{O} \to F_P$ is denoted $f(P)$. In particular, if P is a nonzero point, then $f(P) \in k$, and the pair of functions (x,y) define a map $\phi(P) := (x(P), y(P))$ from the nonzero points of K to points of the affine plane k^2 over k whose coordinates satisfy (2.3.2). The image of the map ϕ is called an *elliptic curve*. We will develop

[3] See Exercises 2.8 and 3.10

some machinery for studying such maps in Chapter 4, but we can illustrate some of the important geometric ideas here without any machinery.

Suppose P_1, P_2, and P_3 are three nonzero points of K with $P_1 \oplus P_2 \oplus P_3 = 0$. What this really means is that $P_1 + P_2 + P_3 - 3P_0 = [z]$ for some principal divisor $[z]$. Then z has a unique pole at P_0 of order 3, and it vanishes at P_1, P_2, and P_3. But $\{1, x, y\}$ is a k-basis for $L(3P_0)$, so there exist constants a, b, c such that $z = a + bx + cy$. Thus,

$$a + bx(P_i) + cy(P_i) = 0 \quad \text{for } i = 1, 2, 3,$$

and we see that $\phi(P_i)$ lies on the line $a + bx + cy = 0$ for $i = 1, 2, 3$. In other words, any three nonzero points of K that sum to zero under $\oplus$ have colinear images in k^2.

Next, suppose a function of the form $z = a + bx + cy \in K$ vanishes at two distinct nonzero points P_1, P_2. In the affine plane we are drawing a line through two points on the curve. If $c \neq 0$, then $z \in L(3P_0) \setminus L(2P_0)$, so we must have $[z] = -3P_0 + P_1 + P_2 + P_3$ for some uniquely determined third point P_3 (not necessarily distinct from P_1 or P_2). Thus z vanishes at a uniquely determined third point, and the three points sum to zero in the Jacobian. In the affine plane, we see that a nonvertical line through any two distinct points of the curve meets the curve at a unique third point.

If $c = 0$, then $b \neq 0$ and $x(P_1) = x(P_2) = -a/b$. Furthermore, $z \in L(2P_0)$, so we must have $[z] = -2P_0 + P_1 + P_2$ and we see that the two points sum to zero in the Jacobian. Conversely, if two points sum to zero in the Jacobian, we get a function $z \in L(2P_0)$ vanishing at those points. In the affine plane, this means that the vertical line drawn through any point on the curve meets the curve at a unique second point, namely its additive inverse under the group law.

Finally, if $z = a + bx + cy$ and $v_P(z) > 1$ for some point $P \in \mathbb{P}_K$, we say that the line $\ell : a + bx + cy = 0$ is *tangent*[4] to the curve at the point $\phi(P)$. If ℓ is vertical, i.e., $z \in L(2P_0)$, we get $2P = 0$. If ℓ is not vertical, i.e., $z \in L(3P_0) \setminus L(2P_0)$, it meets the curve at a third point $Q = -2P$.

Thus, we have the following geometric description of the group law:

Theorem 2.3.3. *Let K be a function field of genus one with at least one point P_0. Then the map ϕ of* (2.2.12) *is bijective. Moreover, if we choose a basis $\{1, x, y\}$ for $L(3P_0)$ as in* (2.3.1) *and embed the nonzero points of K into k^2 via the map $P \mapsto (x(P), y(P))$, then three points of K sum to zero in the Jacobian if and only if their images are collinear in the (x, y)-plane, and two points of K sum to zero in the Jacobian if and only if their x-coordinates are equal.* □

[4]We discuss tangents in detail in Chapter 4.

2.4 Geometric Function Fields

At this point in our exposition we want to apply the theory of derivatives and differential forms developed in Section 1.3. To do so, however, we need an additional hypothesis in order to deal with some difficulties that may arise when the ground field is not perfect. See Exercise 2.10 for an example.

Definition. We say that a function field K/k is *geometric* if $k' \otimes_k K$ is a field for every finite extension k' of k.

Equivalently, we could say that if a field K' contains K and a finite extension k' of k, then k' and K are linearly disjoint over k. Thus, (A.0.11) immediately gives

Lemma 2.4.1. *If K/k is a geometric function field and K_0 is an intermediate field transcendental over k, then K_0/k is also a geometric function field.* □

One way (in fact, as we will prove, the only way) to construct a geometric function field is to let $f(X,Y) \in k[X,Y]$ be an irreducible polynomial that remains irreducible in $k'[X,Y]$ for any finite extension k' of k. Such a polynomial is called *absolutely irreducible*. Since f is irreducible, it generates a prime ideal of $k[X,Y]$, and the quotient ring $k[X,Y]/(f)$ is therefore an integral domain. If we put $x := X + (f)$ and $y := Y + (f)$, we see that the field of fractions $K := k(x,y)$ of $k[X,Y]/(f)$ is a finite extension of $k(x)$. Moreover, for any finite extension k' of k, $k' \otimes_k K = k'(x,y)$, where we identify x, y with $1 \otimes x, 1 \otimes y$ respectively. Since x and y satisfy the irreducible polynomial $f(X,Y)$ over k', it follows that $k'[x,y] \simeq k'[X,Y]/(f)$ is an integral domain, and thus $k'(x,y)$ is a field. To see that K/k is a geometric function field, it only remains to show that k is algebraically closed in K, but this follows from

Lemma 2.4.2. *Let $k \subseteq K$ be fields such that $k' \otimes_k K$ is a field for every finite extension k' of k. Then k is algebraically closed in K.*

Proof. If k is not algebraically closed in K, there is a finite simple extension $k' = k(u)$ for some $u \in K \setminus k$. Since k' is a direct summand of K as a k-vector space, $k' \otimes_k K$ contains the finite dimensional subalgebra $A := k' \otimes_k k'$, and it suffices to show that A is not an integral domain. This is a basic fact, the point being that if it were an integral domain, finite-dimensionality would force every subring of the form $k[v]$ to be a field for all $v \in K^\times$, which would imply that A itself is a field. However, there is a nontrivial homomorphism of A onto k' mapping $u \otimes 1 - 1 \otimes u$ to zero. □

Corollary 2.4.3. *Let k be a field and let $f(X,Y) \in k[X,Y]$ be absolutely irreducible. Then the field of fractions of $k[X,Y]/(f)$ is a geometric function field over k.* □

Corollary 2.4.4. *Let K/k be a geometric function field and let k' be a finite extension of K. Then $k' \otimes_k K$ is a geometric function field over k'.*

Proof. For any finite extension k'' of k' we have

$$k'' \otimes_k' (k' \otimes_k K) = (k'' \otimes_k' k') \otimes_k K = k'' \otimes_k K$$

by associativity of the tensor product, and the result follows from (2.4.2). □

The converse of (2.4.2) is in general false, as shown by Exercise 2.10, but it is true if the ground field is perfect.

Lemma 2.4.5. *Let K be a function field over a perfect ground field k. Then K is geometric.*

Proof. Let k' be a finite extension of k. Then k'/k is separable, so $k' = k(u)$ for some $u \in k'$ by (A.0.17). Moreover, u satisfies an irreducible separable polynomial $f(X) \in k[X]$ of degree $n = |k' : k|$. We claim that f remains irreducible over K. Namely, any factor f_0 has roots that are are algebraic over k, but since the coefficients are symmetric functions of the roots, the coefficients of f_0 are also algebraic over k. Since K/k is a function field, we see that $f_0(X) \in k[X]$ and thus that $f_0 = f$.

Now identify k' and K with their canonical images in $K' := k' \otimes K$. Since u satisfies an irreducible polynomial of degree n over K, $K[u]$ is a field and $|K[u] : K| = n = \dim_K K'$. We conclude that $K' = K(u)$ is a field, as required. □

Recall that the construction of Hasse derivatives given in section 1.3 requires the existence of separating variables. This is automatic in characteristic zero, but in positive characteristic a basic fact about geometric function fields is that they contain separating variables. In fact, we next prove that all $x \in K$ that are not separating variables lie in a unique subfield of index p generated by k and the image of the pth power map. We denote this subfield by kK^p

Theorem 2.4.6. *Let K/k be a geometric function field of characteristic $p > 0$. Then $|K : kK^p| = p$, and the following statements are equivalent for an element $x \in K$:*

1. *$x \in kK^p$.*

2. *$d_{K/k}x = 0$.*

3. *x is not a separating variable for K/k.*

Proof. It is obvious that any k-derivation of K vanishes on kK^p, so 1) implies 2). From 2) we deduce 3) by (1.3.6).

To show that 3) implies 1), choose $x \in K \setminus k$ such that $K/k(x)$ is inseparable. Then (A.0.9) yields a subfield $E \subseteq K$ containing $k(x)$ with K/E purely inseparable of degree p. Then $kK^p \subseteq E$, and $K = E(y)$ for some $y \in K$ with $a := y^p \in E$. If $a \in kE^p$, we can write

$$a = \sum_{i=1}^{r} \alpha_i b_i^p,$$

where $\alpha_i \in k$ and $b_i \in E$ for all i. Let k' be the finite extension of k obtained by adjoining the pth roots of $\alpha_1, \ldots, \alpha_r$, and put $K' = k' \otimes_k K$. By (2.4.4) K'/k' is a geometric function field. Let $\beta_i \in k'$ with $\beta_i^p = \alpha_i$, and define

$$y' = \sum_{i=1}^{r} \beta_i \otimes b_i \in k'E.$$

Then $y'^p = a$. Since the polynomial $X^p - a$ has at most one root in any field of characteristic p, we get $y = y' \in k'E$ and then $K' = k'K = k'E(y) \subseteq k'E$. However, $k'E$ and K are linearly disjoint over E by (A.0.11), a contradiction that shows that $a \notin kE^p$.

Since E is a geometric function field by (2.4.1), we may assume, by induction on $|K : k(x)|$, that $E/k(a)$ is separable, and we conclude from (1.3.5) that every nonzero derivation of E is nonzero at a. However, since $a = y^p$, we have $d_{K/k}(a) = 0$, and therefore $d_{K/k}$ vanishes on E. Since $\ker d_{K/k}$ is a subfield and E is a maximal subfield, either $E = \ker d_{K/k}$ or $\Omega_{K/k} = 0$. The latter case is impossible because the formal derivative on $E[X]$ vanishes on $X^p - a$ and therefore defines a nonzero derivation δ on $E[X]/(X^p - a) \simeq K$ given by

$$\delta(\sum_{i=0}^{p-1} a_i y^i) = \sum_{i=1}^{p-1} i a_i y^{i-1},$$

for $a_i \in E$. It follows that $E = \ker(d_{K/k})$ is unique, and therefore contains all elements $x \in K$ for which $K/k(x)$ is inseparable. Since E also contains kK^p, it only remains to show that $|K : kK^p| \leq p$.

Let $x \in K \setminus E$. Then $K/k(x)$ is separable, so we can choose $y \in K$ with $K = k(x,y)$ by (A.0.17), and we have $K^p = k^p(x^p, y^p)$. Consider the tower

$$kK^p = k(x^p, y^p) \subseteq k(x, y^p) \subseteq K.$$

Since $K/k(x)$ is separable and K/kK^p is purely inseparable, we conclude that $K = k(x, y^p) = kK^p(x)$. Since x is a root of $X^p - x^p$ over kK^p, we have $|K : kK^p| \leq p$, as required. □

From (1.3.6) we immediately get

Corollary 2.4.7. *Let K/k be a geometric function field. Then* $\dim_K \Omega_K = 1$. □

Corollary 2.4.8. *K/k is a geometric function field if and only if $K = k(x,y)$ where x and y satisfy an absolutely irreducible polynomial $f(X,Y) \in k[X,Y]$, in which case $k[x,y] \simeq k[X,Y]/(f)$.*

Proof. We already have one implication from (2.4.3). Conversely, suppose that K/k is geometric, and choose a separating variable x by (2.4.6). Then $K = k(x,y)$ for some $y \in K$ by (A.0.17), where y satisfies an irreducible polynomial of degree $n := |K : k(x)|$ with coefficients in $k(x)$. Carefully clearing denominators, we obtain a polynomial $f(X,Y) \in k[X,Y]$ with $f(x,y) = 0$ such that if we write

$$f(X,Y) = \sum_{i=0}^{n} a_i(X)Y^i$$

with $a_n(X) \neq 0$, then the polynomials $a_i(X) \in k[X]$ are relatively prime. We claim that $f(X,Y)$ is absolutely irreducible. Namely, let k' be a finite extension of k and put $K' := k' \otimes_k K$, with k' and K identified with their natural images in K'. Then $K' = k'(x,y)$ is a field by hypothesis and k' and K are linearly disjoint over k. By (A.0.11) $k'(x)$ and K are linearly disjoint over $k(x)$ and $|K' : k'(x)| = |K : k(x)| = n$. In particular, y cannot satisfy a polynomial of degree less than n over $k'(x)$. If $f(X,Y) = g(X,Y)h(X,Y)$ over $k'[X,Y]$, then one of the factors, say g, must be a polynomial in X alone, but then $g(X) \mid a_i(X)$ for all i and therefore $g(X)$ is a constant since the $a_i(X)$ are relatively prime.

It remains to show that the kernel of the obvious map $k[X,Y] \to k[x,y]$ is (f), or in other words, that x and y satisfy no further relations. Let K' be the field of fractions of the integral domain $k[X,Y]/(f) = k[x',y']$, where $x' = X + (f)$ and $y' = Y + (f)$. There is a map $\phi : k[x',y'] \to k[x,y]$ mapping (x',y') to (x,y) because $f(x,y) = 0$, and ϕ restricts to the obvious isomorphism $\phi_0 : k[x'] \to k[x]$. But ϕ_0 has a unique extension to $k(x')$ and then to an isomorphism $\phi_0^* : K' \to K$ mapping y' to y by elementary field theory. Since ϕ_0^* and ϕ agree on x' and y', they agree on $k[x',y']$. □

A serious problem that arises for nonperfect ground fields is that there may be a prime divisor $P \in \mathbb{P}_K$ for which the residue field F_P is an inseparable extension of k. In such a case, for example, we can't use (1.2.14) to expand elements of $\hat{K}_P$ as Laurent series in powers of a local parameter. We will call a prime divisor P of K/k a *separable prime divisor* if F_P/k is separable, and we denote the set of separable prime divisors by $\mathbb{P}_K^{\text{sep}}$. We will call an arbitrary divisor separable if each of its prime divisors is separable. For any prime divisor P, we denote by F_P^{sep} the maximal separable subextension of F_P/k.

The following result allows us to at least construct infinitely many separable primes.

Theorem 2.4.9. *Let K'/K be a finite separable extension of function fields with $P' \in \mathbb{P}_{K'}$, and put $P := P' \cap K$. Then for almost all P', $e(P'|P) = 1$ and $F_{P'}/F_P$ is separable.*

Proof. By (A.0.17), $K' = K(u)$ for some element $u \in K'$. Let $f(X) := X^n + a_1X^{n-1} + \cdots + a_n$ be the minimum polynomial of u over K. Then $f(X)$ has distinct roots $u = u_1, u_2, \ldots, u_n$ in some extension field of K'. Let $\Delta = \prod_{i<j}(u_i - u_j)$. Then Δ^2 is a symmetric function of the u_i and is therefore a polynomial in the a_i. In particular, $\Delta^2 \in K$.

Now for almost all prime divisors P of K, we have $\nu_P(\Delta^2) = \nu_P(a_i) = 0$ for all i. For any such prime P, (1.1.23) applies, because $f(X)$ has coefficients in $\mathcal{O}_P$ and distinct roots modulo P, and the theorem follows. □

We apply (2.4.9) to the extension $K/k(x)$, using (2.4.6) to choose a separating element $x \in K$. We see that for almost all separable irreducible polynomials $f(x) \in k[x]$, every prime divisor P of f in K is separable because F_P/F separable, where $F := k[x]/(f(x))$ is a separable extension of k. Recall that in order to get any inseparable extensions at all, k must be infinite, since finite fields are perfect. In particular, therefore, all prime divisors of $x-a$ are separable, for almost all $a \in k$, and we have

Corollary 2.4.10. *A geometric function field has infinitely many separable prime divisors.* □

Although the problem of inseparable residue field extensions is a serious one, the problem of inseparable extensions of the function field itself is essentially confined to the corresponding problem for the ground field, in the following sense.

Lemma 2.4.11. *Let K/k be a geometric function field of characteristic $p > 0$. Then kK^p is the unique subfield of K containing k for which K/K_0 is purely inseparable of degree p. If K_0/k_0 is any geometric subfield of K of finite index, then the natural map $\Omega_{K_0/k_0} \to \Omega_{K/k}$ is zero if $K_0 \subseteq kK^p$ and is an embedding otherwise.*

Proof. If K/K_0 is purely inseparable of degree p, then $K^p \subseteq K_0$, and since $|K : kK^p| = p$ by (2.4.6), the first assertion follows. For any geometric $K_0 \subseteq K$, the natural map is either zero or an embedding, because $\dim_{K_0} \Omega_{K_0} = 1$ by (2.4.7). But it is nonzero if and only if $K_0 \not\subseteq kK^p$ by (2.4.6). □

We say that a finite extension K'/K of geometric function fields is *weakly separable* if $K \not\subseteq k'K'^p$, or equivalently, if the natural map $\Omega_K \to \Omega_{K'}$ is an embedding. The main point about this definition is that every weakly separable extension is obtained by first making a (possibly inseparable) constant field extension followed by a separable extension.

Lemma 2.4.12. *Let K'/k' be a weakly separable finite extension of K/k. Then $K'/k'K$ is separable.*

Proof. We may assume that $\mathrm{char}(K) =: p > 0$. Since k' is the full field of constants of K', it is also the full field of constants of $k'K$, so replacing K by $k'K$, we may as well assume that $k' = k$. But if K'/K is inseparable, there is a subfield K_0 containing K with K'/K_0 purely inseparable of degree p, contrary to (2.4.11). □

2.5 Residues and Duality

In this section we study the structure of the module of differential forms on a function field K. Put $\Omega_K := \Omega_{K/k}$ and $d := d_{K/k}$. We will apply Tate's theory of residues from Section 1.4, obtaining a number of results. We first prove the "residue theorem" that the sum of the local residues of any differential form is zero. Then we use Tate's residue form to obtain a canonical isomorphism between the module of differential forms Ω_K and the module of Weil differentials W_K for geometric function fields. Indeed, for the remainder of the book, most of the facts about differential forms that we obtain require this extra assumption, but recall that it is automatically satisfied for perfect ground fields by (2.4.5).

The first step is to define the local residue map. We recall the notation and terminology of section 1.4. The main point, which is almost trivial, is that for any $P \in \mathbb{P}_K$, $\hat{\mathscr{O}}_P$ is a near $\hat{K}_P$-submodule. Namely, for any $x \in \hat{K}_P$ we have $x\hat{\mathscr{O}}_P \subseteq t^{-i}\hat{\mathscr{O}}_P$ for some $i \geq 0$. Since multiplication by t^i induces an isomorphism $t^{-i}\hat{\mathscr{O}}_P/\hat{\mathscr{O}}_P \simeq \hat{\mathscr{O}}_P/t^i\hat{\mathscr{O}}_P$, we see that

$$\dim_k x\hat{\mathscr{O}}_P/\hat{\mathscr{O}}_P \leq \dim_k \hat{\mathscr{O}}_P/t^i\hat{\mathscr{O}}_P = i\dim_k \hat{\mathscr{O}}_P/t\hat{\mathscr{O}}_P < \infty.$$

We therefore have the local residue map $\operatorname{Res}_P(udv) := \langle u, v\rangle_{\hat{K}_P,\hat{\mathscr{O}}_P}$ defined for all $u, v \in \hat{K}_P$, although we are most interested in its restriction to K. See (2.5.3) below.

Lemma 2.5.1. *If $u, v \in \hat{\mathscr{O}}_P$, then $\operatorname{Res}_P(udv) = 0$. In particular, if α is an adele and $v \in K$, then $\operatorname{Res}_P(\alpha_P dv) = 0$ for almost all $P \in \mathbb{P}_K$.*

Proof. The first statement is immediate from (1.4.9) because $\hat{\mathscr{O}}_P$ is invariant under u and v. The second follows because α and v have only finitely many poles. □

Theorem 2.5.2 (Tate). *Let K/k be a function field, let $S \subseteq \mathbb{P}_K$, and let $\omega \in \Omega_K$. Define*

$$\mathscr{O}_S := \cap_{P \in S} \mathscr{O}_P.$$

Then $\mathscr{O}_S$ is a near K-submodule of K, and

$$\sum_{P \in S} \operatorname{Res}_P(\omega) = \operatorname{Res}^K_{\mathscr{O}_S}(\omega).$$

Proof. Note that the sum is finite by (2.5.1). We will apply the Tate residue theory to the K-submodule A_S of the adele ring A defined by

$$A_S := \{\alpha \in A \mid \alpha_P = 0 \text{ for } P \notin S\}.$$

Put $A_S(D) := A_S \cap A(D)$ for any divisor D, and write $\operatorname{Res}_W := \operatorname{Res}^{A_S}_W$ for W a near K-submodule of A_S. Let $\pi : A \to A_S$ be the natural projection map and put $K_S := \pi(K)$. Note that π is a map of K-modules and restricts to isomorphisms $\pi|_K : K \simeq K_S$, and $\pi|_{\mathscr{O}_S} : \mathscr{O}_S \simeq K_S \cap A_S(0)$.

We first observe that $A_S(0)$ is a near K-submodule of A_S. Namely, if $x \in K$ and $\alpha \in A_S(0)$, then $v_P(x\alpha) \geq v_P(x)$ for all $P \in S$, whence $xA_S(0) \subseteq A_S([x]_\infty)$, and therefore

$$\dim_k(xA_S(0) + A_S(0))/A_S(0) \leq \deg[x]_\infty$$

by (2.1.6). Since $\pi|_K$ is an isomorphism, we have

$$\operatorname{Res}^K_{\mathscr{O}_S} = \operatorname{Res}^{K_S}_{K_S \cap A_S(0)} = \operatorname{Res}_{K_S \cap A_S(0)},$$

because we can enlarge K_S to A_S by (1.4.10).

Now we can apply (1.4.11) to obtain

$$\mathrm{Res}_{K_S+A_S(0)} = \mathrm{Res}_{K_S} + \mathrm{Res}_{A_S(0)} - \mathrm{Res}_{\mathscr{O}_S}.$$

However, K_S is K-invariant and $K_S + A_S(0) = \pi(K + A_K(0))$ and therefore has finite codimension by (2.2.1). Using (1.4.9) and (1.4.10) we conclude that

$$(*) \qquad \mathrm{Res}_{A_S(0)} = \mathrm{Res}^K_{\mathscr{O}_S}.$$

For future reference, we record the special case of $(*)$ in which $S = \{P\}$ for some prime $P \in \mathbb{P}_K$:

$$(2.5.3) \qquad \mathrm{Res}_P = \mathrm{Res}^{\hat{K}_P}_{\hat{\mathscr{O}}_P} = \mathrm{Res}^K_{\mathscr{O}_P},$$

where we are henceforth using the embedding $\hat{K}_P \hookrightarrow A$ to identify $x \in \hat{K}_P$ with the adele that is zero at all primes except P and equals x at P.

Now choose $\omega \in \Omega_K$ and write $\omega = ydx$ for some $x, y \in K$. Let $P_1, \ldots P_n$ be all the primes $P \in S$ where either x or y has a pole and T be the set of all other primes in S. Then we can write

$$(**) \qquad A_S(0) = A_T(0) \bigoplus_{i=1}^{n} \hat{O}_{P_i}.$$

Note that $xA_T(0) + yA_T(0) \subseteq A_T(0)$ because neither x nor y has any poles in S. Then $\mathrm{Res}_P(\omega) = 0$ for all $P \in T$ by (2.5.1), and $\mathrm{Res}_{A_T(0)}(\omega) = 0$ by (1.4.9). Now $(*)$ together with repeated application of (1.4.11) to $(**)$ yields

$$\mathrm{Res}_{\mathscr{O}_S}(\omega) = \mathrm{Res}_{A(0)}(\omega) = \sum_{i=1}^{n} \mathrm{Res}_{P_i}(\omega) = \sum_{P \in S} \mathrm{Res}_P(\omega). \quad \square$$

Corollary 2.5.4 (Residue Theorem). *Let K/k be a function field and let $\omega \in \Omega_K$. Then*

$$\sum_{P \in \mathbb{P}_K} \mathrm{Res}_P(\omega) = 0.$$

Proof. If we take $S = \mathbb{P}_K$ in the theorem, we get $\mathscr{O}_S = k \sim 0$, and the result follows by (1.4.10). $\square$

Even in the case $K = k(x)$, the residue theorem is nontrivial, as Exercise 2.12 shows.

For the remainder of this chapter we impose the extra hypothesis that K/k is a geometric function field. With the residue theorem proved, we are now in a position to justify the term "Weil differential." Let $\omega \in \Omega_K$ and $\alpha \in A_K$. Using (2.5.1), it is clear that the function $\omega^* : A_K \to k$ given by

$$(2.5.5) \qquad \omega^*(\alpha) := \sum_{P \in \mathbb{P}_K} \mathrm{Res}_P(\alpha_P \omega)$$

is well-defined. We claim that ω^* is in fact a Weil differential. To see this, write $\omega = udv$ for some $u,v \in K$. If $P \in \mathbb{P}_K$ and u,v,α are all finite at P, then $\mathrm{Res}_P(\alpha\omega) = 0$ by (1.4.9). In fact, if we put $D := [u]_\infty + [v]_\infty$ and take $\alpha \in A_K(-D)$ it is clear that $\mathrm{Res}_P(\alpha\omega) = 0$ for all P. Thus, ω^* vanishes on $A_K(-D)$. Since it also vanishes on K by (2.5.4), it is a Weil differential.

It is easy to see from the definitions that the duality map $* : \Omega_K \to W_K$ is K-linear. Choose a prime $P \in \mathbb{P}_K$ with local parameter t, and for every element $x \in \mathcal{O}_P$, let $\alpha(x)$ be the adele for which $\alpha_P = t^{-1}x$ and $\alpha_Q = 0$ for $Q \neq P$. Then (1.4.13) yields

$$dt^*(\alpha(x)) = \mathrm{Res}_P(t^{-1}xdt) = \mathrm{tr}_{\mathcal{O}_P/P}(x). \tag{2.5.6}$$

To show that the duality map is nonzero, we take $P \in \mathbb{P}_K^{\mathrm{sep}}$ using (2.4.10). Then $\mathrm{tr}_{F_P} \neq 0$ by (A.0.8), so we can find $x \in \mathcal{O}_P$ with $dt^*(\alpha(x)) \neq 0$ by (2.5.6). Now we have a nonzero K-linear map between two one-dimensional K-vector spaces (see (2.4.7) and (2.2.8)) which is therefore an isomorphism. Moreover, we have $dt^* \neq 0$, whence $dt \neq 0$, and t is a separating variable by (2.4.6). Identify $\hat{K}_P$ with its natural image in the adele ring. Since dt^* vanishes on $\hat{\mathcal{O}}_P$ by (1.4.9) and does not vanish on $t^{-1}\mathcal{O}_P$ by (2.5.6), we have $v_P(dt^*) = 0$ by (2.2.4).

Finally, if P is inseparable, then (2.5.6) shows that dt^* vanishes on $t^{-1}\hat{\mathcal{O}}_P$, whence $v_P(dt^*) \geq 1$. Summarizing, we have proved

Theorem 2.5.7. *Let K be a geometric function field. The map $\omega \mapsto \omega^*$ is a K-linear isomorphism $\Omega_K \to W_K$. Moreover, if $P \in \mathbb{P}_K$ and t is a local parameter at P, then $v_P(dt^*) = 0$ if and only if P is separable, in which case t is a separating variable.* □

We see that for $P \in \mathbb{P}_K^{\mathrm{sep}}$, all local paramters at P have nonzero differentials. What about inseparable primes? It is trivial to construct counterexamples; If $K := k(x)$ and $p(x)$ is an irreducible inseparable polynomial, then p is a local parameter at $k[x]_{(p)}$ and $dp = 0$.

We can now extend valuations on K to the module of differential forms by defining

$$v_P(\omega) := v_P(\omega^*) \quad \text{for all } P \in \mathbb{P}_K,$$
$$[\omega] := \sum_P v_P(\omega)P.$$

It follows from (2.5.7) that to compute $v_P(\omega)$ for $P \in \mathbb{P}_K^{\mathrm{sep}}$, we can choose a local parameter t at P, write $\omega = xdt$ for some $x \in K$, and we have $v_P(\omega) = v_P(x)$. It is not yet clear what to do for P inseparable.

The following properties are immediate consequences of the definition and (2.2.7).

Corollary 2.5.8. *Let K be a geometric function field with $\omega \in \Omega_K$ and $P \in \mathbb{P}_K$. Then,*

$$v_P(x\omega) = v_P(x) + v_P(\omega),$$
$$v_P(\omega + \omega') \geq \min\{v_P(\omega), v_P(\omega')\}.$$

In particular, $[\omega]$ is a canonical divisor and $\deg[\omega] = 2g-2$ *for all $\omega \in \Omega_K$.* $\square$

One consequence of (2.5.7) is that the local residue form cannot vanish. This provides some information about the structure of the completion $\hat{\mathscr{O}}_P$ in the inseparable case.

Corollary 2.5.9. *Let K be a geometric function field, let $P \in \mathbb{P}_K$, and let $x \in K$ be a separating variable. Then* $\mathrm{Res}_P(y\,dx) \neq 0$ *for some $y \in \hat{K}_P$. If F^{sep} is the maximal separable subfield of F_P, then F^{sep} is the maximal finite extension of k contained in $\hat{K}_P$.*

Proof. Since $v_P(dx)$ is finite, there must be an adele α with $\mathrm{Res}_P(\alpha_P dx) \neq 0$, proving the first statement. For the second, we note that any finite extension of k contained in $\hat{K}_P$ lies in $\hat{\mathscr{O}}_P$ by (1.1.7), and is therefore a subfield of F_P. Let k'/k be the maximal subextension of F_P/k contained in $\hat{\mathscr{O}}_P$. Since F^{sep} lifts to a subfield of $\hat{\mathscr{O}}_P$ by (1.2.12), it suffices to show that k'/k is separable, but this follows from (1.4.16), (A.0.8), and the nonvanishing of the residue form. $\square$

The extension of valuations to differential forms provides Ω_K with some interesting additional structure. Although they are not functions, we can now speak of the zeros and poles of differential forms. Let $\Omega_K(D)$ be the inverse image of $W_K(D)$ under the duality isomorphism. Then $\Omega_K(0)$ consists of forms ω with $v_P(\omega) \geq 0$ for all P. Such forms are called *regular* differential forms, or *holomorphic* in the case $k = \mathbb{C}$. We then have the following elegant characterization of the genus, which is often taken to be the definition:

Corollary 2.5.10. *Let K be a geometric function field of genus g, and let $\Omega_K(0)$ denote the space of regular differential forms on K. Then* $\dim_k \Omega_K(0) = g$.

Proof. This is immediate from (2.5.7) and (2.2.2). $\square$

More generally, we get an interesting interpretation of the "error term" in the Riemann–Roch theorem. Namely, for any canonical divisor C and any divisor D we have

$$\dim L(C-D) = g - \delta(D) = \dim W_K(D)$$

by (2.2.9) and (2.2.1), so we have the following restatement of (2.2.9).

Corollary 2.5.11. *Let K be geometric of genus g and let D be a divisor on K. Then*

$$\dim L_K(D) = \deg D + 1 - g + \dim \Omega_K(D). \quad \square$$

This formulation has some additional punch. For example, suppose that D is a nonnegative divisor of degree less than g. Then $\delta(D) < g$, so D is special, and we have

Corollary 2.5.12. *If K is a function field of genus g and D is a nonnegative divisor on K of degree less than g, then* $\dim_k \Omega_K(D) \geq g - \delta(D) > 0$. □

We turn now to the problem of actually computing $\operatorname{Res}_P(\omega)$. We can reduce this in general to the computation of the trace of a matrix by (1.4.12), but in the case that P is separable, there is an elegant answer which we now discuss.

We begin by choosing a local parameter t at $P \in \mathbb{P}_K^{\text{sep}}$ and using (1.2.14) to identify $\hat{\mathcal{O}}_P$, the completion of $\mathcal{O}_P$ at P, with the ring $F_P[[t]]$ of formal power series in t with coefficients in F_P.

Define the "obvious" map $\tilde{D}^{(n)} : \hat{\mathcal{O}}_P \to \hat{\mathcal{O}}_P$ via

$$\tilde{D}^{(n)}\Big(\sum_{m=0}^{\infty} a_m t^m\Big) := \sum_{m=n}^{\infty} \binom{m}{n} a_m t^{m-n}.$$

We do not yet know that $\tilde{D}^{(n)}(\mathcal{O}_P) \subseteq \mathcal{O}_P$, but in any case we get a generalized derivation

$$\tilde{D} : \hat{\mathcal{O}}_P \to \hat{\mathcal{O}}_P[[s]],$$

where s is an indeterminate, because it is straightforward to verify that the $\tilde{D}^{(n)}$ satisfy the product rule. (See the discussion immediately preceeding (1.3.8).) Because it is an embedding and $\hat{\mathcal{O}}_P$ is an integral domain, $\tilde{D}$ extends uniquely to a generalized derivation on the field of fractions $\hat{K}_P$. Then by restriction we have an embedding

$$\tilde{D} : K \to \hat{K}_P[[s]]$$

that is the identity in degree zero and agrees with the Hasse derivative D_t on $k(t)$. Note that D_t is defined because t is a separating variable by (2.5.7). Now (1.3.11) yields $\tilde{D} = D_t$, and we have proved

Theorem 2.5.13. *Suppose that K is geometric and $P \in P_K^{\text{sep}}$ with local parameter t. If the power series expansion of $x \in \mathcal{O}_P$ at P is*

$$x = \sum_{m=0}^{\infty} a_m t^m,$$

with $a_m \in F_P$, then the power series expansion of $D_t^{(n)}(x)$ at P is

$$D_t^{(n)}(x) := \sum_{m=n}^{\infty} \binom{m}{n} a_m t^{m-n}.$$

In particular, $D_t^{(n)}(\mathcal{O}_P) \subseteq \mathcal{O}_P$. □

Using this result and the formula (1.3.10), the Hasse derivatives can be explicitly computed from the Laurent series for any $x \in K$. Moreover, we have

Corollary 2.5.14 (Taylor's Theorem). *Suppose that $P \in P_K^{\text{sep}}$ with local parameter t and $x \in \mathscr{O}_P$. Then*

$$x = \sum_{n=0}^{\infty} D_t^{(n)}(x)(P)t^n.$$

Proof. This follows from (2.5.13) by observing that the constant term in the power series expansion of $D_t^{(n)}(x)$ at t is a_n, the coefficient of t^n in the expansion of x. □

Now we can explicitly compute the local residue map.

Lemma 2.5.15. *Let K/k be a geometric function field and let $P \in \mathbb{P}_K^{\text{sep}}$ with local parameter t and residue field F. Let $u, v \in K$. Then*

$$\text{Res}_P(udv) = \text{tr}_{F/k}(a_{-1}),$$

where a_{-1} is the coefficient of t^{-1} in the Laurent series expansion of $u(dv/dt)$ with respect to t.

Proof. Put $x := u(dv/dt)$, so that $udv = xdt$, and use (1.2.14) to write

$$x = \sum_{i=-n}^{\infty} a_i t^i,$$

with $a_i \in F$. We need to show that $\text{Res}_P(xdt) = \text{tr}_{F/k}(a_{-1})$. Since $F \subseteq \hat{K}_P$ and $\hat{\mathscr{O}}_P$ is F-invariant, we can use (1.4.16) to write

$$\text{Res}_P(xdt) = \text{tr}_{F/k}(\text{Res}'_P(xdt)),$$

where Res'_P is the F-linear residue form defined by computing traces with respect to F rather than k. Now put

$$x_0 := \sum_{i=-n}^{-2} a_i t^i \quad \text{and} \quad x_1 := \sum_{i=0}^{\infty} a_i t^i.$$

Then $x = x_0 + x_1 + a_{-1}t^{-1}$, and since $x_1\mathscr{O} \subseteq \mathscr{O}$ and $t\mathscr{O} \subseteq \mathscr{O}$, (1.4.9) yields $\text{Res}'_P(x_1 dt) = 0$, and from (1.4.12) we get $\text{Res}'_P(x_0 dt) = 0$, and

$$\text{Res}'_P(xdt) = \text{Res}'_P(a_{-1}t^{-1}dt) = \text{tr}_{\hat{\mathscr{O}}_P/t\hat{\mathscr{O}}_P}(a_{-1}) = a_{-1},$$

because $\hat{\mathscr{O}}_P/t\hat{\mathscr{O}}_P = F$.

2.6 Exercises

Exercise 2.1. Let $K := k(X)$ and let D be any nonnegative divisor. Prove directly that $\dim(L(D)) = \deg D + 1$.

Exercise 2.2. Prove the assertion of the text that

$$k[X] = \cup_{n\geq 0} L_K(nP_\infty).$$

Exercise 2.3. Let K be a function field of genus g. Assume that K has a divisor of degree g. Show that K has a nonnegative divisor D_0 of degree g, and that every element of the Jacobian can be represented in the divisor group as a difference $D - D_0$ where D is also nonnegative of degree g.

Exercise 2.4. Let K/k have genus g. Suppose that $S := \{P_1, P_2, \ldots, P_n\}$ is a set of distinct points of K (prime divisors of degree 1), that P_0 is another point of K not in S, and that $f_1, f_2, \ldots, f_m$ are a basis for $L_K(dP_0)$, where $n > d \geq 2g-1$.

(i) Show that $m = d - g + 1$.

(ii) Show that the f_i define linearly independent k-valued functions on S.

(iii) Let V be the n-dimensional k-vector space of k-valued functions on S, and let $L \subseteq V$ be the m-dimensional subspace spanned by the f_i. Let $W := L^\perp$ be the $(n-m)$-dimensional subspace of the dual space V^* which annihilates L. V^* has an obvious basis that can be naturally identified with S. Show that every element of W has at least $d - 2g + 1$ nonzero coordinates with respect to this basis. This fact is the basis for the construction of Goppa codes (see[17]). [Hint: Consider the spaces $L(dP_0 - D)$, where D is a sum of at most $d - 2g + 1$ distinct points of S.]

Exercise 2.5. Let K be an elliptic function field, and let P and Q be distinct points of K.

(i) Find an element $x \in K$ with $[x]_\infty = P + Q$.

(ii) Show that $\mathrm{Gal}(K/k(x)) = \langle \sigma \rangle$ for some automorphism σ of K of order two.

(iii) Show that σ interchanges P and Q. [Hint: Find $y \in K$ with $v_P(y) = 1$ and $v_Q(y) = 0$.]

Exercise 2.6. Let k be algebraically closed and suppose that K/k is an elliptic function field. Choose a point $P_0 \in \mathbb{P}_K$ and let $\oplus$ be the addition rule on P_K defined by embedding $\mathbb{P}_K$ into $J(K)$ using P_0 as base point.

(i) Show that for every $P \in \mathbb{P}_K$ there exists an automorphism σ_P of K such that

$$\sigma_P(x)(Q) = x(P \oplus Q)$$

for every $x \in K$ and every $Q \in \mathbb{P}_K$. [Hint: Use (2.3.3).]

(ii) Show that $P \mapsto \sigma_P$ is a homomorphism of groups.

(iii) Suppose $\oplus'$ is the addition rule corresponding to the base point P_0', and that for each $P \in \mathbb{P}_K$, σ_P' is the corresponding automorphism of K. Show that

$$\sigma_P' = \sigma_P \circ \sigma_{P_0}'.$$

Exercise 2.7. Let $J \subseteq \mathrm{Aut}(k(X))$ be the subgroup generated by $X \mapsto 1/X$ and $X \mapsto 1 - X$. Show that J is isomorphic to the symmetric group on three letters, and the fixed subfield of J is $k(j)$, where

$$j(X) := \frac{(X^2 - X + 1)^3}{X^2(X-1)^2}.$$

[Hint: See (A.0.13).] If k is algebraically closed, show that the map $\lambda \mapsto j(\lambda)$ defines a surjection of $k \setminus \{0,1\}$ onto k.

Exercise 2.8. Let k be algebraically closed with $\mathrm{char}(k) \neq 2$, and let K/k be an elliptic function field.

(i) Show that there exists $y, x \in K$ such that $K = k(x,y)$ and (2.3.2) simplifies to $y^2 = x(x-1)(x-\lambda)$ for some $\lambda \in k \setminus \{0,1\}$. [Hint: Complete the square and then change variables.]

(ii) Show that there exists a uniquely determined point $P \in \mathbb{P}_K$ such that $x \in L(2P)$ and $y \in L(3P)$.

(iii) Suppose that there also exists $\tilde{x}, \tilde{y} \in K$ and $\tilde{\lambda} \in k \setminus \{0,1\}$ such that $K = k(\tilde{x}, \tilde{y})$ and $\tilde{y}^2 = \tilde{x}(\tilde{x}-1)(\tilde{x}-\tilde{\lambda})$. Show that there exists $\tau \in \mathrm{Aut}(K)$ with $\tau(\tilde{x}) = ax + b$ for some $a, b \in k$. [Hint: Exercise 2.5.]

(iv) Show that if $u \in K$ and u^2 is a cubic polynomial in $k[x]$, then $u = cy$ for some $c \in k$.

(v) Argue that $\tilde{\lambda} = \sigma(\lambda)$ for some $\sigma \in J$, where J is the group of permutations of Exercise 2.7. Conclude that the map $K \mapsto j(\lambda)$ establishes a well-defined bijection between isomorphism classes of elliptic function fields K/k and elements of k.

Exercise 2.9. Let K be a function field of genus 2. Show that $K = k(x,y)$ where $|K : k(x)| = 2$ and $|K : k(y)| = 5$.

Exercise 2.10. Let k_0 be a perfect field of characteristic $p > 2$ and let $k := k(s,t)$ where s and t are indeterminates. Let $K = k(x,y)$ where $y^p = s + tx^p$.

(i) Show that there are k-derivations δ_x, δ_y of K into K such that $\delta_x(x) = \delta_y(y) = 1$ and $\delta_x(y) = \delta_y(x) = 0$. [Hint: K is a purely inseparable extension of degree p over two different subfields.]

(ii) Show that $\dim_K \Omega_{K/k} = 2$. Conclude that K has no separating variables.

(iii) Let $k' := k(s^{1/p}, t^{1/p})$. Show that $k' \otimes_k K$ is not a field.

(iv) Show that k is algebraically closed in K. [Hint: Use the basis $\{1, y, \ldots, y^{p-1}\}$ for $K/k(x)$.]

Exercise 2.11. Let $f(t)$ be a rational function of t that is finite at $t = 0$. Show that

$$\mathrm{Res}_{t=0} \frac{f(t)}{t} = f(0),$$
$$\mathrm{Res}_{t=0} \frac{f(t)}{t^2} = f'(0).$$

Exercise 2.12. (E.W. Howe[12]) Suppose that $\mathrm{char}(k) \neq 2$, and consider the function

$$g(x) := \prod_{i=1}^{n} \frac{x + a_i}{x - a_i},$$

where the a_i are distinct nonzero elements of k. Use Exercise 2.11 to show that

$$\mathrm{Res}_{x=a_j} g(x)dx = 2a_j \prod_{i:i \neq j} \frac{a_j + a_i}{a_j - a_i}$$

and

$$\mathrm{Res}_{x=\infty} g(x)dx = -2 \sum_{i=1}^{n} a_i.$$

Obtain the identity

$$\sum_{j=1}^{n} a_j \prod_{i:i \neq j} \frac{a_j + a_i}{a_j - a_i} = \sum_{j=1}^{n} a_j.$$

3

Finite Extensions

In this chapter we consider a pair of function fields $K' \supseteq K$ with $|K' : K| < \infty$. Recall that for a function field K/k, the ground field k is algebraically closed in K by definition. However, it may happen that there are additional elements in K' algebraic over k. Denoting the set of all such elements by k', we will often say that K'/k' is a finite extension of K/k. If K is geometric, which we will be assuming throughout the chapter, then $k'K \simeq k' \otimes_k K$, and we have

$$|k' : k| = |k'K : K| \leq |K' : K|.$$

Let $Q \in \mathbb{P}_{K'}$. Since the residue field F_Q is a finite extension of k' and therefore also of k, v_Q cannot vanish on K. This implies that $\mathcal{O}_Q \cap K$ is a valuation ring of K with prime ideal $P := Q \cap K$. In this situation we will say that Q *divides* P, or sometimes that Q *lies above* P.

We want to apply the results of the last chapter on differential forms, which require that K/k be geometric. Here is the basic fact we need to know:

Lemma 3.0.1. *Let K/k be a geometric function field and let K'/k' be a finite extension of K/k. Then K'/k' is geometric.*

Proof. Since K/k is geometric, $k' \otimes_k K$ is a field isomorphic to the subfield $k'K \subseteq K'$. By (2.4.4) $k'K/k'$ is geometric, so we may as well assume that $k' = k$.

Now let $k \subseteq k_0 \subseteq k_1$ with k_1/k finite, and put $K'_i := k_i \otimes_k K'$. We need to show that K'_1 is a field. If the inclusions are proper, then K'_0 and $k_1 \otimes_{k_0} K'_0 \simeq K'_1$ are geometric by induction on $|k_1 : k|$.

We are therefore reduced to the case $k_1 = k(\alpha)$ for some $\alpha \in k_1$ with minimum polynomial $f(X)$ over k. Identifying k_1 and K' with their natural images in K'_1 as usual, we have $K'_1 = K'[\alpha]$, so it suffices to show that f is irreducible over K'.

However, all roots of f are algebraic over k, so the coefficients of any factor of f over K' are elements of K' algebraic over k and therefore lie in k. Thus, f remains irreducible over K', and therefore K'_1 is a field. □

The hypothesis that K is geometric will be assumed throughout the chapter.

3.1 Norm and Conorm

Recall that if P (resp. P') is a prime of K (resp. K') with $P'|P$, we denote the ramification index (resp. the residue degree) of P' over P by $e(P'|P)$ (resp. $f(P'|P)$). Given a finite extension K'/k' of K/k, we define two homomorphisms: $N_{K'/K} : \mathrm{Div}(K') \to \mathrm{Div}(K)$ and $N^*_{K'/K} : \mathrm{Div}(K) \to \mathrm{Div}(K')$ called the norm and conorm, respectively. For each prime Q of K', let $P := Q \cap K$, put

$$N_{K'/K}(Q) := f(Q|P)P,$$

and extend linearly to $\mathrm{Div}(K')$. For each prime P of K, let

$$N^*_{K'/K}(P) := \sum_{Q|P} e(Q|P)Q,$$

and extend linearly to $\mathrm{Div}(K)$. We first record an easy consequence of the multiplicativity of e and f in a chain of extensions (1.1.25):

Lemma 3.1.1. *Suppose that $K \subseteq K' \subseteq K''$ are function fields. Then*

$$N_{K'/K} \circ N_{K''/K'} = N_{K''/K}, \quad \textit{and}$$
$$N^*_{K''/K'} \circ N^*_{K'/K} = N^*_{K''/K}. \quad \square$$

Recall that the degree of a prime divisor P is the degree of the residue field $\mathcal{O}_P/P$ over the constant field k. So we have to be careful when computing degrees to take into account a possible constant field extension. If $x \in K \subseteq K'$ we will use the notation $[x]_K$ (resp. $[x]_{K'}$) to denote the principal divisor of x in K (resp. K').

Lemma 3.1.2. *Let K'/k' be a finite extension of K/k, let $D' \in \mathrm{Div}(K')$, and let $D \in \mathrm{Div}(K)$. Then*

$$\deg N_{K'/K}(D') = |k' : k| \deg D',$$
$$\deg N^*_{K'/K}(D) = \frac{|K' : K|}{|k' : k|} \deg D,$$
$$N_{K'/K}(N^*_{K'/K}(D)) = |K' : K|D, \quad \textit{and}$$
$$[x]_{K'} = N^*_{K'/K}([x]_K) \quad \textit{for all } x \in K.$$

Proof. Let P be a prime divisor of K, and let $\{Q_1, \ldots, Q_r\}$ be the set of all distinct prime divisors of P in K'. Put $e_i := e(Q_i|P)$ and $f_i := f(Q_i|P)$ for all i. In addition, let F be the residue field $\mathcal{O}_P/P$ and let $F_i := \mathcal{O}_{Q_i}/Q_i$. By linearity we may assume that $D = P$ and $D' = Q_1$. Then

$$\deg Q_i = |F_i : k'| = \frac{|F_i : k|}{|k' : k|} = \frac{|F_i : F||F : k|}{|k' : k|} = \frac{f_i \deg P}{|k' : k|},$$

and the first formula follows. Moreover, (2.1.17) yields

$$\deg N^*(P) = \sum_{i=1}^{r} e_i \deg Q_i = \sum_{i=1}^{r} \frac{e_i f_i}{|k' : k|} \deg P = \frac{|K' : K|}{|k' : k|} \deg P.$$

Similarly, we obtain

$$N(N^*(P)) = \sum_{i=1}^{r} e_i f_i P = |K' : K| P.$$

The last formula is immediate from the definition of the conorm. □

As an immediate consequence, it follows that the conorm is an injective map $\mathrm{Div}(K) \hookrightarrow \mathrm{Div}(K')$. Using this map, we can identify $\mathrm{Div}(K)$ with a subgroup of $\mathrm{Div}(K')$. From the last formula above, this makes sense on principal divisors, and identifies $[x]_K$ with $[x]_{K'}$. For this reason, some authors call the conorm the "inclusion map."

We turn now to the nontrivial result of this section, namely the induced maps on Jacobians. It is easy to see that the conorm maps principal divisors to principal divisors and therefore induces a homomorphism $N^*_{K'/K} : J_K \to J_{K'}$. Less obvious is the fact that the norm also maps principal divisors to principal divisors, so there is also an induced homomorphism $N_{K'/K} : J_{K'} \to J_K$. In fact, more is true: The norm is really an extension of the ordinary field norm of (A.0.2) to the divisor group.

Theorem 3.1.3. *Let K' be a finite extension of K, and let $[x]_K$ (resp. $[x]_{K'}$) denote the principal divisor of x in K (resp. K'.) Then*

$$N_{K'/K}([x]_{K'}) = [N_{K'/K}(x)]_K$$

for all $x \in K'$, where $N_{K'/K}(x)$ is the field norm of (A.0.2).

Proof. Suppose that $x = yz$. Dropping subscripts, we have $N(x) = N(y)N(z)$ and $N([x]) = N([y]) + N([z])$. Therefore, if we can prove the formula for y and z, it follows for x. We will refer to this property of the formula as "linearity."

The formula is equivalent to the following statement: Let P be a prime of K and let $x \in K'$. Then

$$v_P(N_{K'/K}(x)) = \sum_{Q|P} f(Q|P) v_Q(x). \tag{3.1.4}$$

To prove this, we first suppose that $x \in K$. Put $n := |K' : K|$. Then $N_{K'/K}(x) = x^n$, and (3.1.4) follows from (2.1.17). Next, we reduce to the case that x is integral over $\mathcal{O}_P$. Namely, clearing denominators of the coefficients of the minimum polynomial, we have

$$\sum_{i=1}^{m} r_i x^i = 0$$

for some integer m, with $r_i \in \mathscr{O}_P$. Multiplying through by r_m^{m-1}, we see that $r_m x$ is integral over $\mathscr{O}_P$. Since (3.1.4) holds for r_m, it holds for x if and only if it holds for $r_m x$ by linearity.

Let R be the integral closure of $\mathscr{O}_P$ in K'. Then R is a free $\mathscr{O}_P$-module of rank n by (2.1.18) and (1.1.9). For $x \in R$, $N(x) = \det M_x$, where M_x is the matrix of multiplication by x with respect to any K-basis of K'. Since an $\mathscr{O}_P$-basis of R is a K-basis of K', we have $N(R) \subseteq \mathscr{O}_P$.

For each prime $Q|P$ there is, by the weak approximation theorem, an element $t_Q \in K'$ with $v_{Q'}(t_Q) = \delta_{QQ'}$ for all $Q'|P$. Then $t_Q \in R$ by (1.1.8), t_Q is a local parameter at Q, and we have

$$x = u \prod_{Q|P} t_Q^{v_Q(x)},$$

where $u \in R$ is a unit. By linearity, it suffices to prove (3.1.4) for $x = u$ or $x = t_Q$. However, when $x \in R^\times$, we have $N(x) \in \mathscr{O}_P^\times$, and both sides of (3.1.4) are zero. So we may finally assume that $x = t_Q$ for some $Q|P$. In this case, (3.1.4) reduces to

$$v_P(N(t_Q)) = f(Q|P).$$

The columns of M_x span the free submodule xR of R. By (1.1.12) and (1.1.13), there are nonnegative integers $e_1, e_2, \ldots, e_s$ such that

$$R/xR \simeq \bigoplus_{i=1}^{s} \mathscr{O}_P/t^{e_i}\mathscr{O}_P$$

and therefore $v_P(\det M_x) = \sum_i e_i$. It follows that $v_P(N(x))$ is the length of the finite $\mathscr{O}_P$-module R/xR. But for $x = t_Q$, $R/t_Q R \simeq \mathscr{O}_Q/Q$ by (1.1.22). So the length of $R/t_Q R$ as an $\mathscr{O}_P$-module is just the dimension of $\mathscr{O}_Q/Q$ over $\mathscr{O}_P/P$, which is the residue degree $f(Q|P)$ as required. □

Given a finite separable extension K'/K, there is a trace map $\Omega_{K'} \to \Omega_K$ which we now define. Let x be a separating variable for K (see (2.4.6)). Then $d_{K'}(x) \neq 0$ because the extension is weakly separable. Now every $\omega \in \Omega_{K'}$ can be written $\omega = y\,dx$ for some $y \in K'$, and we define

$$\operatorname{tr}_{K'/K}(\omega) := \operatorname{tr}_{K'/K}(y)dx.$$

From the K-linearity of $\operatorname{tr}_{K'/K} : K' \to K$, we easily deduce that $\operatorname{tr}_{K'/K}$ is K-linear on $\Omega_{K'}$ and is independent of the choice of separating variable $x \in K$. Using the trace map, we can get a formula relating residues of a differential form on K' with residues of its trace. For simplicity, we will assume that K' and K have the same constant field k.

Theorem 3.1.5 (Trace Formula). *Let K/k be a geometric function field, let K'/k be a finite separable extension, let $P \in \mathbb{P}_K$, and let $\omega \in \Omega_{K'}$. Then*

$$\operatorname{Res}_P(\operatorname{tr}_{K'/K}(\omega)) = \sum_{Q|P} \operatorname{Res}_Q(\omega).$$

Proof. Let $x \in K$ be a separating variable as above, and put $\omega = ydx$ for some $y \in K'$. If we take S in Tate's theorem (2.5.2) to be the set of prime divisors Q of P in K', we get

$$\operatorname{Res}^{K'}_{\mathscr{O}_S} = \sum_{Q|P} \operatorname{Res}_Q.$$

Since $\mathscr{O}_S$ is the integral closure of $\mathscr{O}_P$ in K' by (1.1.8), it is a finitely generated $\mathscr{O}_P$-module by (2.1.18). Then (1.1.13) applies, and since $\mathscr{O}_S$ is torsion-free, we see that $\mathscr{O}_S$ is in fact free. Because the field of fractions of $\mathscr{O}_S$ is K', it follows that $\mathscr{O}_S$ has rank $n = |K' : K|$. Let $\{x_1, \dots, x_n\}$ be an $\mathscr{O}_P$-basis for $\mathscr{O}_S$. Then $\{x_1, \dots, x_n\}$ is also a K-basis for K', and applying (1.4.14) to the near K-submodule $\mathscr{O}_P \subseteq K$, the result follows. □

3.2 Scalar Extensions

We say that K'/k' is a *scalar extension* of K/k if $K' = k'K$. In this situation we sometimes say that K' is *defined over* k. For any extension K' of K, if k' is the subfield of K' algebraic over k, we can put $L := k'K$ and then think of the extension as consisting of two steps: a scalar extension $L \supseteq K$, followed by an extension $K' \supseteq L$ of fields with the same constant field. Because we are assuming that K is geometric, we have $K' \simeq k' \otimes_k K$ for any scalar extension K' of K, and the following facts are clear:

Lemma 3.2.1. *Let K/k be a geometric function field, let K' be a finite extension of K, and suppose that $K' = k'K$ where k' is a finite extension of k. Then $|k' : k| = |K' : K|$. If, in addition, k'/k is separable, then K'/K is also separable. If $\{x_1, \dots, x_n\} \subseteq K$ is linearly independent over k, it remains linearly independent over k'.* □

The behavior of a geometric function field K/k under a scalar extension k'/k differs markedly depending on whether k'/k is separable or not. In the former case, things work out more or less "as expected," but in the latter case there can be some unpleasant surprises, which we shall postpone until Section 3.4. For now, all we need is

Lemma 3.2.2. *Let $P \in \mathbb{P}_K$ and let k'/k be a finite extension. If either F_P/k or k'/k is purely inseparable, then there is a unique divisor of P in $K' := k' \otimes_k K$.*

Proof. Case 1): k'/k is purely inseparable. Then for some power $q = p^n$ the map $x \mapsto x^q$ is an isomorphism of K' onto a subfield of K. For $x \in K'$ define

$$v'(x) := v_P(x^q).$$

Then v' is a homomorphism of $K'^{\times}$ to an infinite cyclic group satisfying the ultrametric inequality (1.1.1), so it is a discrete valuation of K' that clearly divides v_P. On the other hand, if Q is any divisor of P in K' and $v_Q(x) \geq 0$, then $v_Q(x^q) \geq 0$, and therefore $v_P(x^q) \geq 0$ as well. This implies that $\mathcal{O}_{v'} = \mathcal{O}_Q$, and therefore Q is unique.

Case 2): F_P/k is purely inseparable. Let k'_s/k be the maximal separable subextension of k'/k and put $K'_s := k'_s \otimes_k K$. Let Q be a divisor of P in K'_s. Since F_P and k'_s are linearly disjoint over k by (A.0.10), it follows that $|F_Q : F_P| \geq |k'_s : k|$, whence the inequality is an equality and Q is the unique divisor of P in K'_s by (2.1.17). Since k'/k'_s is purely inseparable, the result now follows from case 1). □

Theorem 3.2.3. *Suppose that K/k is a geometric function field and that k'/k is finite and separable. Put $K' := k' \otimes_k K$ and let $Q \in \mathbb{P}_{K'}$ with $P := Q \cap K$. Then $k' \otimes_k \mathcal{O}_P$ is the integral closure of $\mathcal{O}_P$ in K' and $e(Q|P) = 1$.*

Proof. We have $k' = k(\alpha)$ for some $\alpha \in k'$ by (A.0.17). Let $f(X)$ be the minimal polynomial of α over k. Then $f(X)$ is irreducible over K by (3.2.1). Let R be the integral closure of $\mathcal{O}_P$ in K'. Since $\alpha \in \mathcal{O}_Q$ and all coefficients of $f(X)$ lie in $\mathcal{O}_P$, (1.1.23) says that $e(Q|P) = 1$ and that $\mathcal{O}_P[\alpha] = R$. Since $\mathcal{O}_P[\alpha] \subseteq k' \otimes_k \mathcal{O}_P \subseteq R$, we have proved that $R = k' \otimes_k \mathcal{O}_P$. □

Unfortunately, for inseparable scalar extensions there is some *tsouris*[1] here, as shown by Exercise 3.12. This leads us to the following important definition.

Definition. Let K/k be a geometric function field and let k'/k be a finite extension. Following [18], we say that the prime divisor $P \in \mathbb{P}_K$ is *singular* with respect to k' if $k' \otimes_k \mathcal{O}_P$ is properly contained in the integral closure of $\mathcal{O}_P$ in $k' \otimes_k K$, and nonsingular with respect to k' otherwise. We say that P is singular if it is singular with respect to some finite extension k'/k, and nonsingular otherwise.

Note that if k is perfect, then all prime divisors are nonsingular by (3.2.3). We will defer the study of singular primes to Section 3.4, where we show that there is a finite, purely inseparable scalar extension K'/K such that all prime divisors of K' are nonsingular.

We say that an extension k'/k is a *splitting field* for $P \in \mathbb{P}_K$ if $\deg Q = 1$ for every prime divisor Q of P in $k' \otimes_k K$.

Lemma 3.2.4. *Suppose that $P \in \mathbb{P}_K$ is nonsingular with respect to k'. Put $K' := k' \otimes_k K$, and let $Q \in \mathbb{P}_{K'}$ with $Q|P$. Then $F_Q = k'F_P$. In particular, if P is nonsingular then F_P is a splitting field for P.*

Proof. Let R be the integral closure of $\mathcal{O}_P$ in K'. Then (1.1.22) yields $\mathcal{O}_Q = R + Q$, from which it follows immediately that $F_Q = k'F_P$. In particular, $F_Q = k'$ when $k' \supseteq F_P$. □

---

[1] A yiddish expression meaning trouble. Perhaps a more accurate translation would be "heartburn."

An important point about geometric function fields is that we can extend scalars to the algebraic closure $\overline{k}$ of k, where, for example, the results of Chapter 4 will apply. Thus, for a geometric function field K, we define $\overline{K} := \overline{k} \otimes_k K$. Since $k' \otimes_k K$ is a field for every finite extension k'/k, any embedding $k' \to \overline{k}$ extends to an embedding $k' \otimes_k K \to \overline{K}$. Indeed, $\overline{K}$ is just the set-theoretic union of the images of such embeddings. In particular, every element of $\overline{K}$ lies in some subfield and is therefore invertible, so $\overline{K}$ is a field. Thus, $\overline{K}$ is a (geometric) function field over $\overline{k}$. Summarizing these observations, we have

Lemma 3.2.5. *Let K/k be a geometric function field, let $\overline{k}$ be the algebraic closure of k, and let $\overline{K} := \overline{k} \otimes_k K$. Then $\overline{K}/\overline{k}$ is a function field and if k' is a finite extension of k, every embedding $k' \to \overline{k}$ extends to an embedding $k' \otimes_k K \to \overline{K}$.* □

We next extend valuations from K to $\overline{K}$. Since $\overline{K}/K$ is not of finite degree, some care must be taken, but because $\overline{K}$ is a union of finite scalar extensions, the problem is not serious. Note at the outset that all prime divisors on $\overline{K}$ are points[2] and that if $Q \in \mathbb{P}_{\overline{K}}$ and $P := Q \cap K \in \mathbb{P}_K$ then $e(Q|P)$ is still well-defined via $v_Q(t) = e(Q|P)$ for t a local parameter at P.

Theorem 3.2.6. *Let K/k be a geometric function field and let P be a nonsingular prime divisor of K. Let k'/k be a finite extension that is a splitting field for P, and put $K' := k' \otimes_k K$. Then:*

1. *There are exactly $|F_P^{\text{sep}} : k|$ distinct points P' of K' with $P' \cap K = P$, and for each such point P' we have $e(P'|P) = |F_P : F_P^{\text{sep}}|$.*

2. *There is a one-to-one correspondence between points Q of $\overline{K}$ satisfying $Q \cap K = P$ and points P' of K' dividing P, given by $Q \cap K' = P'$, and $e(Q|P) = e(P'|P)$.*

Proof. 1) Let P' be a divisor of P in $\mathbb{P}_{K'}$. By hypothesis and (3.2.4) we have $k' = F_{P'} = k'F_P$ and therefore $k' \supseteq F_P$. Put $k'_0 := F_P^{\text{sep}}$ and consider the tower

$$K \subseteq K'_0 := k'_0 \otimes_k K \subseteq K'.$$

Let $P_1, \dots, P_r$ be the prime divisors of P in K'_0. By (3.2.3) we have $e(P_i|P) = 1$. Since P is nonsingular, (3.2.4) implies that $F_{P_i} = k'_0 F_P = F_P$, and thus $f(P_i|P) = 1$ for all i. Then (2.1.17) yields $r = |K'_0 : K| = |k'_0 : k|$.

Moreover, F_{P_i}/k'_0 is purely inseparable, so it follows from (3.2.2) that for each i there is a unique prime divisor P'_i of P_i in K'. Since k' is a splitting field for P, $F_{P'_i} = k'$ and thus $f(P'_i|P_i) = |k' : F_P|$ for all i. Now (2.1.17) and (1.1.25) yield

$$e(P'_i|P) = e(P'_i|P_i) = \frac{|K' : K'_0|}{|k' : F_P|} = \frac{|k' : k'_0|}{|k' : F_P|} = |F_P : F_P^{\text{sep}}|.$$

[2] In the language of algebraic geometry, what we are calling the points of $\overline{K}$ would normally be called the points of the (projective nonsingular) curve C defined by K, and what we are calling the points of K, i.e., the prime divisors of K of degree one, would be called the k-rational points of C.

2) By (1.1.6) there is a point Q of $\overline{K}$ with $Q \cap K = P$. Since P splits into a sum of points in K', $P' := Q \cap K'$ must be one of those points. To complete the proof, we argue that there is exactly one point Q' of $\overline{K}$ extending each divisor P' of P in K'. If, by way of contradiction, there were more than one $\overline{k}$-valuation of $\overline{K}$ restricting to a multiple of $v_{P'}$ on K', they would differ on some element $u \in \overline{K}$. If we write

$$u = \sum_{i=1}^{r} \alpha_i \otimes x_i$$

with $\alpha_i \in \overline{k}$ and $x_i \in K$, then $k'' := k'(\alpha_1, \dots, \alpha_r)$ is a finite extension of k', $u \in k'' \otimes_k K$, and we would have more than one prime divisor P'' of the point P' in $k'' \otimes_k K$. However, since $f(P''|P') = |k'' : k'|$, (2.1.17) tells us that there is a unique divisor of P' in K''.

A similar argument shows that a local parameter t at Q lies in some finite extension $K'' := k'' \otimes_k K$, where we may assume that $k'' \supseteq k'$. If we put $P'' := Q \cap K''$, then clearly $e(Q|P'') = 1$, and $e(P''|P') = 1$ by (2.1.17) as above. We conclude that $e(Q|P) = e(Q|P'')e(P''|P')e(P'|P) = e(P'|P)$. □

It is clear from (3.2.6) that for any prime divisor P on K, the conorm map $N^*_{\overline{K}/K}(P)$ is well-defined. If $k' \subseteq \overline{k}$ and $K' := k' \otimes_k K$, we say that the point Q of $\overline{K}$ is *defined over* k' if $Q \cap K'$ is a point, or equivalently if $Q = N^*_{\overline{K}/K'}(P)$ for some divisor P (necessarily a point) on K'.

More generally, we say that a divisor $D' \in \mathrm{Div}(\overline{K})$ is defined over k' if D is in the image of the conorm map $N^*_{\overline{K}/K'}$. We say that k' is a splitting field for a divisor D if it is a splitting field for every prime divisor P with $v_P(D) \neq 0$.

Corollary 3.2.7. *Let D be a divisor on $\overline{K}$. Then D is defined over some finite extension k' of k.*

Proof. We may assume without loss of generality that D is a point Q. Let $P := Q \cap K$, let k' be a splitting field for P, and apply (3.2.6). □

3.3 The Different

In this section we introduce an important invariant of a weakly separable finite extension of geometric function fields K'/k' over K/k. Recall from (2.4.11) that there is a natural identification of $\Omega_{K/k}$ with a K-subspace of $\Omega_{K'/k'}$. So given $\omega \in \Omega_{K/k}$, we have both the divisor $[\omega]_K \in \mathrm{Div}(K)$ as a differential form on K and the divisor $[\omega]_{K'} \in \mathrm{Div}(K')$ as a differential form on K'. The main point of this section is to study the relationship between these two divisors.

Let x be a separating variable in K, let $y \in K$, let $Q \in \mathbb{P}_{K'}$ with $P := Q \cap K$, and put $e := e(Q|P)$. Then it follows from (2.5.8) that the quantity

$$v_Q(ydx) - ev_P(ydx) = v_Q(dx) - ev_P(dx)$$

does not depend on y, and we have

Lemma 3.3.1. *Let K'/k' be a weakly separable finite extension of K/k and let $\omega \in \Omega_{K/k}$. Let $Q \in \mathbb{P}_{K'}$ and put $P := Q \cap K$. Then the integer*

$$d(Q|P) := v_Q(\omega) - e(Q|P)v_P(\omega)$$

is independent of ω and depends only on Q and P. □

We call $d(Q|P)$ the *different exponent* of Q over P. From the definitions, we see that $d(Q|P)$ is the coefficient of Q in the divisor $[\omega]_{K'} - N^*_{K'/K}([\omega]_K)$. In particular, we have $d(Q|P) = 0$ for almost all P and Q, and we define

$$\mathscr{D}_{K'/K} := \sum_{Q \in \mathbb{P}_{K'}} d(Q|P)Q.$$

We call $\mathscr{D}_{K'/K}$ the *different* of the extension. Thus, by definition we have

$$[\omega]_{K'} = N^*_{K'/K}([\omega]_K) + \mathscr{D}_{K'/K}. \quad \square \tag{3.3.2}$$

It is not hard to see how the different behaves in a tower of extensions:

Lemma 3.3.3. *Suppose that $K/k \subseteq K'/k' \subseteq K''/k''$ are function fields with K'' weakly separable over K. Then*

$$\mathscr{D}_{K''/K} = \mathscr{D}_{K''/K'} + N^*_{K''/K'}(\mathscr{D}_{K'/K}).$$

Proof. By (3.3.2) we have

$$[\omega]_{K'} = N^*_{K'/K}([\omega]_K) + \mathscr{D}_{K'/K},$$
$$[\omega]_{K''} = N^*_{K''/K'}([\omega]_{K'}) + \mathscr{D}_{K''/K'},$$
$$[\omega]_{K''} = N^*_{K''/K}([\omega]_K) + \mathscr{D}_{K''/K}.$$

The result follows by taking the conorm $N^*_{K''/K'}$ of the first equation, applying (3.1.1), substituting into the second equation, and then equating with the third. □

Corollary 3.3.4. *Suppose that $K \subseteq K' \subseteq K''$ is a tower of weakly separable finite extensions, and $Q'' \in \mathbb{P}_{K''}$. Put $Q' := Q'' \cap K'$ and $Q := Q'' \cap K$. Then*

$$d(Q''|Q) = d(Q''|Q') + e(Q''|Q')d(Q'|Q).$$

In particular, if any two of the integers $d(Q''|Q), d(Q''|Q'), d(Q'|Q)$ are zero, the third is also zero. □

If we now take degrees in (3.3.2) and use (3.1.2) we obtain

Theorem 3.3.5 (Riemann–Hurwitz). *Let K/k be a geometric function field, and let K'/k' be a finite weakly separable extension of K/k. If g_K (resp. $g_{K'}$) denotes the genus of K (resp. K') then*

$$2g_{K'} - 2 = \frac{|K' : K|}{|k' : k|}(2g_K - 2) + \deg \mathscr{D}_{K'/K}. \quad \square$$

The Riemann–Hurwitz formula is quite important, but it doesn't tell us anything until we know something about the degree of the different, or what is essentially the the same thing, the different exponent. In the case that Q and P are separable, we can compute $d(Q|P)$ as follows.

Choose local parameters s at Q and t at P. Then $t = s^e u$, where $e = e(Q|P)$ and u is a unit. Since Q and P are separable, s and t are separating variables and $\nu_Q(ds) = 0 = \nu_P(dt)$ by (2.5.7), and we have

$$d(Q|P) = \nu_Q(dt) = \nu_Q(es^{e-1}u\,ds + s^e\,du).$$

It follows that $d(Q|P) = e - 1$, provided that $\operatorname{char}(K) \nmid e$. However, when $\operatorname{char}(K)|e$ we have $d(Q|P) \geq e$.

Summarizing, we have

Theorem 3.3.6. *Let K' be a finite weakly separable extension of K and let Q be a separable prime of K' dividing the separable prime P of K. Then $d(Q|P) = e(Q|P) - 1$ unless* $\operatorname{char}(K)|e(Q|P)$, *in which case $d(Q|P) \geq e(Q|P)$.* □

When $\operatorname{char}(K) \mid e(Q|P)$ we say that $Q|P$ is *wildly ramified*; when $e(Q|P) > 1$ and $\operatorname{char}(K) \nmid e(Q|P)$, we say that $Q|P$ is *tamely ramified*. Certainly all ramification is tame in characteristic zero. We will analyze wild ramification further in (3.5.9).

For readers willing to assume that the ground field is perfect, or even better, of characteristic zero, (3.3.6) settles the calculation of $d(Q|P)$ and puts some teeth into the Riemann–Hurwitz formula. For those of us determined to push onward, however, the calculations of (3.3.6) do not work when Q is inseparable, because $\nu_Q(ds) \neq 0$. Instead, our strategy will be to first extend scalars, but this is easier said than done, essentially because P or Q may be singular. See Section 3.4 for the gory details.

Even after extending scalars to a splitting field, a question remains as to how to compute the different exponent, particularly in the wildly ramified case. If local parameters s and t can be explicitly found, $d(Q|P)$ can be obtained by expanding t in powers of s, perhaps by the method of undetermined coefficients, as was illustrated in Section 1.1. However the following result gives a useful alternative.

Theorem 3.3.7. *Let K' be a finite weakly separable extension of K and suppose that Q is a separable prime of K' dividing the separable prime P of K with $f(Q|P) = 1$. If s is a local parameter at Q such that $\nu_{Q'}(s) = 0$ for every prime divisor $Q' \neq Q$ of P in $\mathbb{P}_{K'}$, and $f(X)$ is the (monic) minimum polynomial of s over K, then $K' = K(s)$ and $\nu_Q(f'(s)) = d(Q|P)$.*

Proof. Put $K_1 := K(s)$ and $Q_1 := Q \cap K_1$. Then $e(Q|Q_1) = 1$ and Q is the unique prime divisor of Q_1 in $\mathbb{P}_K$ because s is a local unit at every other prime divisor of P. Since $f(Q|Q_1)$ divides $f(Q|P) = 1$ by (1.1.25), we conclude that $K_1 = K$ by (2.1.17).

By (1.1.8), s is integral over $\mathcal{O}_P$, so $f(X)$ is monic of degree $n := |K' : K|$ with coefficients in $\mathcal{O}_P$. Put $f(X) = X^n + \sum_{i=0}^{n-1} a_i X^i$ with $a_i \in \mathcal{O}_P$. Then $a_0 = N_{K'/K}(s)$, whence our hypothesis and (3.1.4) imply that a_0 is a local parameter at P. Differentiating $f(s) = 0$ we obtain

$$0 = f'(s)ds + da_0 + s\sum_{i=1}^{n-1} da_i s^{i-1},$$

whence

$$v_Q(f'(s)ds) = \min\left\{v_Q(da_0), v_Q\left(s\sum_{i=1}^{n-1} da_i a^{i-1}\right)\right\}.$$

Since a_0 is a local parameter at P, P is separable, and the a_i are all integral, we have $da_i/da_0 \in \mathcal{O}_P$ by (2.5.7). This implies that

$$v_Q(da_0) \leq v_Q\left(\sum_{i=1}^{n-1} da_i s^{i-1}\right) < v_Q\left(s\sum_{i=1}^{n-1} da_i s^{i-1}\right).$$

Since Q is also separable we have $v_Q(ds) = 0$ and therefore

$$v_Q(f'(s)) = v_Q(f'(s)ds) = v_Q(da_0) = d(Q|P). \quad \square$$

We note that whenever $f(Q|P) = 1$ (e.g., when k is algebraically closed!) a local parameter can be found satisfying the hypotheses of (3.3.7) by using the weak approximation theorem (1.1.16). However, in the special case that P is totally ramified in K', the hypotheses are automatically satisfied by any local parameter, and we have

Corollary 3.3.8. *Let K' be a finite weakly separable extension of K and let Q be a separable prime of K' dividing the separable prime P of K. Suppose that $e(Q|P) = |K' : K|$, and let s be a local parameter at Q with minimum polynomial $f(X)$ over K. Then $d(Q|P) = v_Q(f'(s))$.* □

The definition of the different given above is not the standard one, so for the remainder of this section we develop the classical theory following Hecke [11], whose treatment closely follows Dedekind's original one. This material will be mainly used later, in the study of singularities of plane curves. To simplify the exposition, we will deal only with separable prime divisors. Indeed, we will often assume that relevant residue field extensions are trivial. In Section 3.4 we will show that these assumptions always hold in some finite scalar extension.

If $y \in K'$ has monic characteristic polynomial $f(X)$ with respect to K'/K, the quantity $\delta_{K'/K}(y) := f'(y)$ is called the *different* of y over K. When $K' = K(y)$, which we will usually assume, we will just write $\delta_K(y) := \delta_{K(y)/K}(y)$. The reason for this terminology is that if we write

$$f(X) = \prod_{i=1}^{n}(X - y_i)$$

with $y_1 = y$, then it is immediate that

$$f'(y) = \prod_{i=2}^{n} (y - y_i).$$

To understand the role of $\delta_K(y)$, we recall the Lagrange interpolation formula. Let $y_1, \ldots, y_n$ be indeterminates, and consider the polynomials

$$F_k(X) = \sum_{i=1}^{n} y_i^k \prod_{j \neq i} \frac{X - y_j}{y_i - y_j},$$

where $0 \leq k \leq n$. Then $F_k(X)$ is a polynomial in X of degree at most $n-1$ over $k(y_1, \ldots, y_n)$ and $F_k(y_i) = y_i^k$ for $1 \leq i \leq n$. This implies that $F_k(X) = X^k$ for $0 \leq k < n$. For $k = n$, put

$$f(X) := \prod_{i=1}^{n} (X - y_i).$$

Then $F_n(X)$ and $X^n - f(X)$ both have degree at most $n-1$ and agree at each y_i, so they are equal. It follows that

$$\sum_{i=1}^{n} \frac{y_i^k f(X)}{f'(y_i)(X - y_i)} = \begin{cases} X^k & \text{for } 0 \leq k < n, \\ X^n - f(X) & \text{for } k = n. \end{cases}$$

Setting $X = 0$ and dividing by $f(0)$, we obtain

$$\sum_{i=1}^{n} \frac{y_i^k}{f'(y_i)} = \delta_{k,n-1},$$

where δ_{ij} is the Kronecker delta. Specializing the y_i to the roots of a separable irreducible polynomial over K and using (A.0.4), we have

Lemma 3.3.9. *If y is separable over K of degree n, then for $0 \leq k < n$ we have*

$$\operatorname{tr}_{K(y)/K} \frac{y^k}{\delta_K(y)} = \begin{cases} 1 & \text{if } k = n-1, \\ 0 & \text{otherwise.} \end{cases} \qquad \square$$

The above result connects the different with the trace. To explore this connection further, let K'/K be finite and separable, fix a prime $P \in \mathbb{P}_K$, and for any subring $R \subseteq K'$ let R^* denote the dual of R with respect to the trace form:

$$R^* := \{y \in K' \mid \operatorname{tr}_{K'/K}(yR) \subseteq \mathcal{O}_P\}.$$

Then (3.3.9) says that $\delta_K(y)^{-1} \in \mathcal{O}_P[y]^*$, provided that $K' = K(y)$. However, more is true. Denote by R_P the integral closure of $\mathcal{O}_P$ in K'. The ring R_P was studied in (1.1.22). By (2.1.18) it is a finitely generated $\mathcal{O}_P$-module.

Lemma 3.3.10. *Suppose that $K' = K(y)$, where y is separable over K and integral over $\mathcal{O}_P$ for some $P \in \mathbb{P}_K$. Then $\mathcal{O}_P[y]^* \delta_K(y) = \mathcal{O}_P[y]$. Furthermore, $R_P^* \delta_K(y)$ is the unique largest ideal of R_P contained in $\mathcal{O}_P[y]$.*

Proof. Because y is integral, $\mathcal{O}_P[y] \subseteq \mathcal{O}_P[y]^*$ and $\mathcal{O}_P[y]^*$ is an $\mathcal{O}_P[y]$-module. We therefore get $\delta_K(y)^{-1}\mathcal{O}_P[y] \subseteq \mathcal{O}_P[y]^*$ by (3.3.9). To obtain the reverse inclusion let $x \in \mathcal{O}_P[y]^*$. Then we have

$$\delta_K(y)x = \sum_{i=0}^{n-1} a_i y^i$$

for some $a_i \in K$. If, by way of contradiction, $a_i \notin \mathcal{O}_P$ for some i, choose r maximal with this property and write

$$\delta_K(y)xy^{n-1-r} = \sum_{i=0}^{r} a_i y^{i+n-1-r} + w,$$

where $w \in \mathcal{O}_P[y]$. Then using (3.3.9) we get

$$a_r = \operatorname{tr}\left(\delta_K(y)^{-1}\sum_{i=0}^{r} a_i y^{i+n-1-r}\right) = \operatorname{tr}\left(\delta_K(y)^{-1}w - xy^{n-1-r}\right) \in \mathcal{O}_P,$$

a contradiction that shows that $\delta_K(y)\mathcal{O}_P[y]^* \subseteq \mathcal{O}_P[y]$.

Finally, let

$$C_P(y) := R_P^*\delta_K(y) \subseteq \mathcal{O}_P[y].$$

Evidently, $C_P(y)$ is an ideal of R_P. Let $I \subseteq \mathcal{O}_P[y]$ be any ideal of R_P. Then

$$R_P\delta_K(y)^{-1}I \subseteq I\delta_K(y)^{-1} \subseteq \mathcal{O}_P[y]^*.$$

In particular, $\operatorname{tr}(R_P\delta_K(y)^{-1}I) \subseteq \mathcal{O}_P$ and therefore $\delta_K(y)^{-1}I \subseteq R_P^*$, or equivalently, $I \subseteq C_P(y)$ as required. □

In general, whenever $R_1 \subseteq R_2$ are rings, there is a unique largest ideal of R_2 (possibly the unit ideal) contained in R_1, which can be described as the annihilator of the R_1-module R_2/R_1. This ideal is called the *conductor* of R_1 in R_2. Thus, (3.3.10) says that $C_P(y)$ is the conductor of $\mathcal{O}_P[y]$ in R_P.

We can now obtain another characterization of the different exponent.

Theorem 3.3.11. *Suppose that $P \in \mathbb{P}_K^{\text{sep}}$, K'/K is a finite weakly separable extension, and that $Q \in \mathbb{P}_{K'}^{\text{sep}}$ divides P with $f(Q|P) = 1$. Then*

$$\min_{x \in R_P^*} v_Q(x) = -d(Q|P).$$

Proof. By the weak approximation theorem (1.1.16) there exists $s \in K'$ such that $v_Q(s) = 1$ and $v_{Q'}(s-1) > 0$ for all prime divisors $Q' \neq Q$ of P. Then s satisfies the hypotheses of (3.3.7), from which we obtain $v_Q(\delta_K(s)) = d(Q|P)$. For any $x \in R_P^*$ (3.3.10) yields $x\delta_K(s) \in C_P(s) \subseteq R_P$ and therefore $v_Q(x) \geq -d(Q|P)$. Moreover, we get equality for some x if we can find a Q-local unit in $C_P(s)$, because $R_P^*\delta_K(s) = C_P(s)$.

Let $N_{K'/K}(\delta_K(1-s)) =: ut^k$, where t is a local parameter at P and u is a P-local unit. Put $v := (1-s)^k$ and $e := \max_{Q'|P} e(Q'|P)$. Then uv^e is a Q-local unit in $\mathcal{O}_P[s]$. We will show that in fact $uv^e \in C_P(s)$. Let $x \in R_P$. Then we need to show that $xuv^e \in \mathcal{O}_P[s]$.

We have $\mathcal{O}_P + Q = \mathcal{O}_Q$ because $f(Q|P) = 1$. Multiplying by s^i yields

$$\mathcal{O}_P s^i + Q^{i+1} = \mathcal{O}_P s^i + Qs^i = \mathcal{O}_Q s^i = Q^i,$$

from which it follows that

$$\mathcal{O}_P[s] + Q^n = \mathcal{O}_Q$$

for any integer $n > 0$. In particular, there exists $x' \in \mathcal{O}_P[s]$ such that

$$w := x - x' \in Q^{ek}.$$

It therefore suffices to show that $wuv^e \in \mathcal{O}_P[s]$. We have

$$\frac{wuv^e}{\delta_K(1-s)} = \frac{wuv^e}{\delta_K(1-s)} \cdot \frac{N(\delta_K(1-s))}{ut^k} = \frac{N(\delta_K(1-s))}{\delta_K(1-s)} \cdot \frac{wv^e}{t^k}.$$

We claim that the right-hand side lies in R_P. Namely, for any $y \in R_P$, $N(y)/y$ is a product of conjugates of y, each of which is integral over $\mathcal{O}_P$, whence $N(y)/y \in R_P$. For all divisors $Q' \neq Q$ of P we have $v_{Q'}(1-s) \geq 1$, so our choice of e guarantees that $v_{Q'}(v^e) \geq v_{Q'}(t^k)$. Since $v_Q(w) \geq ek$, we conclude that

$$\frac{wuv^e}{\delta_K(1-s)} \in R_P,$$

and therefore

$$wuv^e \in R_P \delta_K(1-s) \subseteq C_P(1-s) \subseteq \mathcal{O}_P[1-s] = \mathcal{O}_P[s]. \quad \square$$

From (3.3.11) we can see why R_P^* is sometimes called the (local) *inverse different*. We remark that the hypothesis $f(Q|P) = 1$ in (3.3.11) can be weakened to the requirement that the residue field extension F_Q/F_P be separable. See Exercise 3.5 for details.

Given a prime divisor $Q \in \mathbb{P}_{K'}$ of $P \in \mathbb{P}_K$ and $y \in K'$, let $\mathcal{O}_P[y]_Q$ denote the localization of $\mathcal{O}_P[y]$ at the prime ideal $Q \cap \mathcal{O}_P[y]$. The main result we need for applications to plane curves is the following.

Corollary 3.3.12. *Suppose $P \in \mathbb{P}_K^{\text{sep}}$, $K' = K(y)$ for some separable element $y \in K'$ that is integral over $\mathcal{O}_P$, and $Q \in \mathbb{P}_{K'}^{\text{sep}}$ divides P with $f(Q|P) = 1$. Then*

$$v_Q(\delta_K(y)) \geq d(Q|P),$$

and if equality holds, then $\mathcal{O}_P[y]_Q = \mathcal{O}_Q$.

Proof. Put $M := Q \cap \mathcal{O}_P[y]$. From (3.3.10) we have $R_P^* \delta_K(y) = C_P(y)$ and the inequality is immediate from (3.3.11). Moreover, equality holds precisely when $C_P(y) \not\subseteq M$.

Let $\{Q = Q_1, \ldots, Q_r\}$ be the set of all prime divisors of P in K', and let $\mathscr{V} := \{v_1, \ldots, v_r\}$ be the corresponding set of valuations. Then, in the notation of (1.1.17), $C_P(y) = K'(\mathscr{V}; e)$ for some nonnegative function $e : \mathscr{V} \to \mathbb{Z}$.

Suppose now that $C_P(y) \not\subseteq M$. Then $e(v_1) = 0$. We certainly have $\mathcal{O}_P[y]_Q \subseteq \mathcal{O}_Q$. Conversely, choose any $x \in \mathcal{O}_Q$, then there exists $b \in K'$ with $v_1(b) = 0$ and

$$v_i(b) \geq e(v_i) + |v_i(x)| \quad (2 \leq i \leq r),$$

by the weak approximation theorem (1.1.16) . This implies that $b \in C_P(y) \setminus M$ and $xb \in C_P(y)$. We conclude that $x \in \mathcal{O}_P[y]_Q$ as required. □

3.4 Singular Prime Divisors

In this section, we fill a gap in the previous section relating to the computation of the different for inseparable primes. Namely, a certain singularity condition for prime divisors was identified, which, when present, made it difficult to extend scalars in a natural way. In this section we prove that for any geometric function field K/k, at most finitely many primes of K are singular and there is a finite, purely inseparable scalar extension K'/K such that all prime divisors of K' are nonsingular. In addition, we show that $\mathscr{D}_{K'/K} \leq 0$, with equality if and only if $K' = K$. This implies that $\mathscr{D}_{K''/K'} = 0$ for all scalar extensions K'' of K', and that $\mathscr{D}_{K''/K'} \geq 0$ for all finite extensions K''/K'. Obviously, this section can be safely skipped by readers willing to assume that k is perfect.

Recall that ,by definition, $P \in \mathbb{P}_K$ is singular with respect to some finite extension of scalars k'/k if $k' \otimes_k \mathcal{O}_P$ is a proper $\mathcal{O}_P$-submodule of the integral closure of $\mathcal{O}_P$ in $K' := k' \otimes_k K$. Here are some basic facts about this situation.

Lemma 3.4.1. *Let K/k be a geometric function field with $P \in \mathbb{P}_K$ and let k'/k be a finite extension. Let R be the integral closure of $\mathcal{O}_P$ in $K' := k' \otimes_k K$ and let $\tilde{R} := k' \otimes_k \mathcal{O}_P$. Then:*

1. *$\dim_k(R/\tilde{R}) < \infty$, and $\tilde{R}$ contains a nonzero ideal of R.*
2. *If P is nonsingular with respect to k' and $Q \in \mathbb{P}_{K'}$ is a divisor of P, then $F_Q = k'F_P$.*
3. *For any finite extension k''/k', the following conditions are equivalent:*
   (a) P is nonsingular with respect to k'',
   (b) P is nonsingular with respect to k' and Q is nonsingular with respect to k'' for all prime divisors Q of P in K'.

Proof. (1) Since both R and $\tilde{R}$ are free $\mathcal{O}_P$-modules of rank $|k' : k|$ (see (2.1.18) and (1.1.9)), the $\mathcal{O}_P$-module $R/\tilde{R}$ has finite length by (1.1.12) and is therefore finite k-dimensional. This implies that the annihilator, C, of $R/\tilde{R}$ in $\tilde{R}$ is a nonzero ideal of $\tilde{R}$ and therefore $\tilde{R}$ contains the nonzero ideal CR of R.

Note that if Q is a divisor of P in K', then $\mathcal{O}_Q = R + Q$ by (1.1.22). This immediately implies (2).

To prove (3), let $\{Q_1, \ldots, Q_m\}$ be the set of all prime divisors of P in K'. For each i, let $\{Q_{i1}, \ldots, Q_{ij_i}\}$ be the set of all prime divisors of Q_i in $K'' := k'' \otimes_k K$, and put $v_{ij} := v_{Q_{ij}}$. In addition, let $\mathcal{O}_i := \mathcal{O}_{Q_i}$, let R_i be the integral closure of $\mathcal{O}_i$ in K'', and let $\tilde{R}_i := k''\mathcal{O}_i$ for each i.

Suppose first that some Q_i is singular with respect to k'' for some i, say $i = 1$. Since $\tilde{R}_1$ contains a nonzero ideal of R_1, (1.1.17) implies that there are integers $e_1, \ldots, e_{j_1}$ such that if $x \in K''$ and $v_{1j}(x) \geq e_j$ for $j = 1, \ldots, j_1$, then $x \in \tilde{R}_1$. Let $y \in R_1 \setminus \tilde{R}_1$. By the weak approximation theorem (1.1.16) there exists $x \in K''$ such that $v_{kj}(x) \geq 0$ for $k > 1$ and all j, and such that $v_{1j}(y - x) \geq e_j$ for $j = 1, \ldots, j_1$. In particular, $x - y \in \tilde{R}_1$, and x is integral over $\mathcal{O}_P$. We see that $x \notin \tilde{R}_1 \supseteq k''\mathcal{O}_P$ and thus P is singular with respect to k''.

Next, choose a k'-basis $\alpha_1 = 1, \alpha_2, \ldots, \alpha_n$ for k''. Then the α_i are a K'-basis for K'', and every element $x \in K''$ can be uniquely written

$$(*) \qquad x = \sum_{i=1}^{n} \alpha_i \otimes x_i$$

with $x_i \in K'$. If we choose a k-basis $\{\beta_1 = 1, \beta_2, \ldots, \beta_l\}$ for k', then the β_j are a K-basis for K' and we can write

$$x_i = \sum_j \beta_j \otimes x_{ij}$$

with $x_{ij} \in K$. The products $\alpha_i\beta_j$ are a k-basis for k'', so $x \in k'' \otimes_k \mathcal{O}_P$ if and only if $x_{ij} \in \mathcal{O}_P$ for all i, j. Thus, if P is singular with respect to k', we can take $x = x_1 \in R \setminus \tilde{R}$ and conclude that P is singular with respect to k''. This proves that a) implies b).

Moreover, if $x_i \in \tilde{R}$ for all i, then $x \in k'' \otimes_k \mathcal{O}_P$ by associativity of the tensor product. Thus, if P is singular with respect to k'' we can take $x \in K''$ integral over $\mathcal{O}_P$ with $x_i \notin \tilde{R}$ for at least one i, say $i = 1$. If all Q_j are nonsingular with respect to k'', then since x is certainly integral over $\mathcal{O}_{Q_j}$, we have $x \in \tilde{R}_j$ for each j. By uniqueness of the expansion $(*)$, this implies that $x_1 \in \cap_j \mathcal{O}_{Q_j}$. Thus, x_1 is integral over $\mathcal{O}_P$ and therefore P is singular with respect to k'. □

Corollary 3.4.2. *All separable prime divisors are nonsingular.*

Proof. Let k'/k be finite and let k'_s/k be the maximal separable subextension of k'/k. Put $K' := k' \otimes_k K$ and $K'_s := k'_s \otimes_k K$. Let $P \in \mathbb{P}_K^{\text{sep}}$, and let R (resp. R_s) be the integral closure of $\mathcal{O}_P$ in K' (resp. K'_s). Let $\mathcal{S}$ be the set of all prime divisors of P in K'_s. We need to show that $R = k' \otimes_k \mathcal{O}_P$.

From (3.2.3) and (3.2.4) we see that $R_s = k'_s \otimes_k \mathscr{O}_P$, and $F_Q = k'_s F_P$ for every prime $Q \in \mathscr{S}$. In particular, each $Q \in \mathscr{S}$ is separable, so by (3.4.1) we may assume that $k = k_s$, or in other words that k'/k is purely inseparable.

Put $\tilde{R} := k' \otimes_k \mathscr{O}_P$ and let $Q \in \mathbb{P}_{K'}$ be a divisor of P. Since F_P/k is separable by hypothesis, k' and F_P are linearly disjoint over k by (A.0.10). Now we have $F_Q \supseteq k'F_P = k' \otimes_k F_P$ and thus $f(Q|P) \geq |k' : k| = |K' : K|$. By (2.1.17) $f(Q|P) = |K' : K|$ and Q is the unique divisor of P in K'. It follows that $F_Q = k' \otimes_k F_P$ and thus that $\mathscr{O}_Q = \tilde{R} + Q$. Since $\mathscr{O}_Q = R$ is a finitely generated $\mathscr{O}_P$-module by (2.1.18), Nakayama's lemma (1.1.5) yields $\tilde{R} = \mathscr{O}_Q$. □

Because there does not seem to be any easy way to compute $d(Q|P)$ in the inseparable case, we will have to refer back to the definition in what follows. The following lemma is useful for this purpose.

Lemma 3.4.3. *Let k' be a finite extension of k with $K' := k' \otimes_k K$. Let $Q \in \mathbb{P}_{K'}$ and put $P := Q \cap K$. Suppose that the subfields k' and $\hat{K}_P$ of $\hat{K}'_Q$ are linearly disjoint over their intersection $k'_Q := k' \cap \hat{K}_P$. Then for every differential form $\omega \in \Omega_K$ we have*

$$\operatorname{Res}_P(\omega) = \operatorname{tr}_{k'_Q/k} \operatorname{Res}_Q(\omega).$$

Proof. We first apply (1.2.11) to conclude that $\hat{K}'_Q = k'\hat{K}_P$. Thus, our hypothesis says that there is an isomorphism

$$(*) \qquad \hat{K}'_Q \simeq k' \otimes_{k'_Q} \hat{K}_P.$$

This puts us in a position to apply (1.4.15) and (1.4.16). For $x, y \in K$, denote by $\operatorname{Res}'_P(ydx)$ the residue form defined by the near $\hat{K}_P$-submodule $\hat{\mathscr{O}}_P$ with respect to the ground field k'_Q. Then (1.4.16) yields

$$\operatorname{Res}_P(ydx) = \operatorname{tr}_{k'_Q/k} \operatorname{Res}'_P(ydx).$$

By virtue of $(*)$ and (1.4.15), we see that $R := k' \otimes_{k'_Q} \hat{\mathscr{O}}_P$ is a near $\hat{K}'_Q$-module, and that

$$\operatorname{Res}'_P(ydx) = \operatorname{Res}_R^{\hat{K}'_Q}(ydx).$$

However, both R and $\hat{\mathscr{O}}_Q$ are free $\hat{\mathscr{O}}_P$-modules of rank $|\hat{K}'_Q : \hat{K}_P|$ (see (1.2.11)), so $\hat{\mathscr{O}}_Q/R$ has finite length by (1.1.12) and is therefore finite-dimensional. Now (1.4.10) yields

$$\operatorname{Res}'_P(ydx) = \operatorname{Res}_Q(ydx).$$

Combining the above, we have proved the lemma. □

We are now ready to proceed with the computation of $d(Q|P)$ for a scalar extension k'/k. We first deal with the case that k'/k is separable (although P and Q may not be)

Theorem 3.4.4. *Let K/k be a geometric function field and let $P \in \mathbb{P}_K$. Suppose that k'/k is a finite separable extension and $K' := k' \otimes_k K$. Then $d(Q|P) = 0$ for every prime divisor Q of P in $\mathbb{P}_{K'}$.*

Proof. There is a finite separable extension k''/k' such that $k'' \supseteq F_P^{\mathrm{sep}}$, the maximal separable subextension of F_P/k. Let Q be a prime divisor of P in K' and let Q' be a prime divisor of Q in $K'' := k'' \otimes_{k'} K' = k'' \otimes_k K$. Suppose we could prove the theorem for the extension k''/k and also for all primes Q of K' with respect to the extension k''/k'. Then we would have $d(Q'|Q) = d(Q'|P) = 0$ and the result would follow for the extension k'/k by (3.3.4). It therefore suffices to prove the theorem under the additional assumption that $k' \supseteq F_P^{\mathrm{sep}}$.

We next argue that the hypothesis of (3.4.3) holds. Namely, by (3.2.3) and (3.2.4) it follows that $F_Q = k'F_P$ and then (A.0.10) yields $f(Q|P) = |k' : F_P^{\mathrm{sep}}|$. Since $e(Q|P) = 1$ by (3.2.3), we have

$$|\hat{K'}_Q : \hat{K}_P| = f(Q|P) = |k' : F_P^{\mathrm{sep}}|$$

by (1.2.11). On the other hand, $F_P^{\mathrm{sep}} \subseteq \hat{K}_P$ by (1.2.12). If we therefore put $k'_Q := k' \cap \hat{K}_P$, we have $|k' : k'_Q| \le |\hat{K'}_Q : \hat{K}_P|$. Since $\hat{K'}_Q = k'\hat{K}_P$ by (1.2.11), we conclude that $k'_Q = F_P^{\mathrm{sep}}$ and that k' and $\hat{K}_P$ are linearly disjoint over k'_Q. By (3.4.3) we then have

$$(*) \qquad \mathrm{Res}_P(\omega) = \mathrm{tr}_{F_P^{\mathrm{sep}}/k} \mathrm{Res}_Q(\omega)$$

for all $\omega \in \Omega_K$.

Now let $x \in K$ be a separating variable with $v_P(dx) = v$, and let t be a local parameter at P. Put $\omega := t^{-v}dx$ then $v_P(\omega) = 0$. Because $d(Q|P)$ is independent of ω, it suffices to show that $v_Q(\omega) = 0$.

Since $k'F_P = F_Q$ and $e(Q|P) = 1$ by (3.2.3) and (3.2.4), we can take inverse images in $\hat{\mathscr{O}}_Q$ (see (1.2.10)) to obtain

$$\hat{\mathscr{O}}_Q = k'\hat{\mathscr{O}}_P + \hat{P}\hat{\mathscr{O}}_Q.$$

But $\hat{\mathscr{O}}_Q$ is a finitely generated $\hat{\mathscr{O}}_P$-module by (1.2.11), so we have $\hat{\mathscr{O}}_Q = k'\hat{\mathscr{O}}_P$ by Nakayama's Lemma (1.1.5).

Let $\{1 = \alpha_1, \alpha_2, \ldots, \alpha_n\}$ be a k-basis for k', and let $y = \sum_i \alpha_i y_i \in \hat{\mathscr{O}}_Q$, where $y_i \in \hat{\mathscr{O}}_P$. Since Res_Q is k'-linear, we have

$$\mathrm{Res}_Q(y\omega) = \sum_i \alpha_i \mathrm{Res}_Q(y_i\omega).$$

We argue that $\mathrm{Res}_Q(y\omega) = 0$, for if not, we get $\mathrm{Res}_Q(y_i\omega) \neq 0$ for some $y_i \in \hat{\mathscr{O}}_P$. Recalling that $F_P^{\mathrm{sep}} \subseteq \hat{\mathscr{O}}_P$ and that Res_Q is k'-linear, it follows from $(*)$ that for any nonzero element $\alpha \in F_P^{\mathrm{sep}}$, we have

$$\mathrm{tr}_{F_P^{\mathrm{sep}}/k}(\alpha \mathrm{Res}_Q(y_i\omega)) = \mathrm{tr}_{F_P^{\mathrm{sep}}/k}(\mathrm{Res}_Q(\alpha y_i\omega)) = \mathrm{Res}_P(\alpha y_i\omega) = 0,$$

contradicting the nondegeneracy of the trace form (A.0.8). This shows that $v_Q(\omega) \geq 0$. On the other hand, if $\text{Res}_Q(t^{-1}y\omega) = 0$ for all $y \in \hat{\mathscr{O}}_Q$, it would follow from (*) that $\text{Res}_P(t^{-1}y\omega) = 0$ for all $y \in \hat{\mathscr{O}}_P$, contrary to $v_P(\omega) = 0$. Thus, $v_Q(\omega) = 0$ as required. □

We are now ready for the main result of this section.

Theorem 3.4.5. *Let K/k be a geometric function field and let $P \in \mathbb{P}_K$. Suppose that K'/k' is a finite scalar extension of K/k. Then $d(Q|P) \leq 0$ for every prime divisor Q of P in $\mathbb{P}_{K'}$, and equality holds for all $Q|P$ if and only if P is nonsingular with respect to k'.*

Proof. We proceed by induction on $|k' : k|$. Suppose that k_0 is a proper intermediate field and put $K_0 := k_0 \otimes_k K$ and $Q_0 := Q \cap K_0$. The induction hypothesis first yields $d(Q|Q_0) \leq 0$ and $d(Q_0|P) \leq 0$, from which (3.3.4) gives us $d(Q|P) \leq 0$ with equality for all $Q|P$ if and only if $d(Q|Q_0) = 0 = d(Q_0|P)$ for all $Q_0|P$ and all $Q|Q_0$. The induction hypothesis then tells us that the latter condition is equivalent to the two conditions P nonsingular with respect to k_0 and Q_0 nonsingular with respect to k' for all $Q_0|P$. Then (3.4.1) completes the proof.

We are therefore reduced to the special case that there are no proper subfields between k and k'. The case that k'/k is separable was proved in (3.4.4), because in that case, P is nonsingular with respect to k' by (3.2.3). We are therefore left with the case k'/k purely inseparable of degree $p = \text{char}(k)$.

Let $x \in K$ be a separating variable with $v_P(dx) = v$, and let t be a local parameter at P. Put $\omega := t^{-v}dx$, then $v_P(\omega) = 0$. Because $d(Q|P)$ is independent of ω, it suffices to show that $v_Q(\omega) \leq 0$, with equality if and only if P is nonsingular with respect to k'.

We again have $\hat{K'}_Q = k'\hat{K}_P$ by (1.2.11), and $|\hat{K'}_Q : \hat{K}_P| = p = e(Q|P)f(Q|P)$ because Q is the unique prime divisor of P in K' by (3.2.2). It is therefore trivial that k' and $\hat{K}_P$ are linearly disjoint over k, and (3.4.3) gives us in this case

$$(*) \qquad \text{Res}_P(\omega) = Res_Q(\omega).$$

Suppose P is nonsingular with respect to k' and $y \in \hat{O}_Q$. Since $\mathscr{O}_Q = k' \otimes_k \mathscr{O}_P$, it follows that $\hat{\mathscr{O}}_Q = k' \otimes_k \hat{\mathscr{O}}_P$, and we can write

$$y = \sum_{i=-0}^{p-1} \alpha_i y_i,$$

where the α_i are a k-basis for k' and $y_i \in \hat{\mathscr{O}}_P$. Then (*) implies that $\text{Res}_Q(y_i\omega) = 0$ for all i, whence $\text{Res}_Q(y\omega) = 0$ because Res_Q is k'-linear. From the definition we conclude that

(**) If P is nonsingular with respect to k', then $v_Q(\omega) \geq 0$.

The remainder of the argument now depends on whether or not k' is isomorphic to a subfield of F_P. If not, then $F_P \subsetneq k'F_P \subseteq F_Q$, and we get $f(Q|P) = p$ and $e(Q|P) = 1$. In particular, $F_Q = k'F_P$. This means that $\mathcal{O}_Q = k' \otimes_k \mathcal{O}_P + Q = k' \otimes_k \mathcal{O}_P + P\mathcal{O}_Q$. However, $\mathcal{O}_Q$ is a finitely generated $\mathcal{O}_P$-module by (2.1.18), so we get $\mathcal{O}_Q = k' \otimes_k \mathcal{O}_P$ by Nakayama's Lemma (1.1.5), and thus P is nonsingular with respect to k'. So in this case we need to prove that $v_Q(\omega) = 0$.

We already have $v_Q(\omega) \geq 0$. If, by way of contradiction, $\operatorname{Res}_Q(t^{-1}y\omega) = 0$ for all $y \in \hat{\mathcal{O}}_Q$, it would follow from $(*)$ that $\operatorname{Res}_P(t^{-1}y\omega) = 0$ for all $y \in \hat{\mathcal{O}}_P$, contrary to $v_P(\omega) = 0$. But t is a local parameter at Q since $e(Q|P) = 1$, and thus $v_Q(\omega) \leq 0$, as required.

Finally, therefore, we are reduced to the case that k' is isomorphic to a subfield of F_P. Write $k' = k(\beta)$ for some $\beta \in k'$ with $\beta^p = \alpha \in k$. Then there exists $b \in \mathcal{O}_P$ with $b^p = \alpha + a$ for some $a \in P$. Put $s := b - \beta$, then $s^p = a$. Thus,

$$s^{-1} = a^{-1}s^{p-1} = a^{-1}\sum_{i=0}^{p-1}(-1)^i\binom{p-1}{i}b^{p-1-i}\beta^i.$$

Let $y \in \mathcal{O}_P$. Using $(*)$ and k'-linearity, we obtain

$$\operatorname{Res}_Q(s^{-1}y\omega) = \sum_{i=0}^{p-1}(-1)^i\binom{p-1}{i}\operatorname{Res}_P(a^{-1}b^{p-1-i}y\omega)\beta^i.$$

When this sum vanishes for a particular y, we get a dependence relation on the β^i over k, and therefore each term of the sum vanishes. Taking the last term in particular, we find that when $\operatorname{Res}_Q(s^{-1}y\omega) = 0$, we have $\operatorname{Res}_P(a^{-1}y\omega) = 0$ as well.

Let $v_Q(s) = n$ and put $e := e(Q|P)$ and $f := f(Q|P)$. Then

$$ev_P(a) = v_Q(a) = v_Q(s^p) = pn = efn,$$

and thus $v_P(a) = fn$.

There exists an element $y \in t^{fn-1}\mathcal{O}_P$ such that $\operatorname{Res}_P(a^{-1}y\omega) \neq 0$, because $v_P(\omega) = 0$. For this y, we have $\operatorname{Res}_Q(s^{-1}y\omega) \neq 0$, and the definition of $v_Q(\omega)$ implies that

$$v_Q(\omega) < -v_Q(s^{-1}y) = n - e(fn-1) = n - pn + e.$$

Since $n \geq 1$, $p \geq 2$, and $e = 1$ or p, we have $n - pn + e \leq 1$, with equality if and only if $n = 1$ and $e = p$. This shows that $v_Q(\omega) \leq 0$ in all cases. From $(**)$ we conclude that $v_Q(\omega) = 0$ if P is nonsingular with respect to k'. Conversely, if $v_Q(\omega) = 0$, then in particular we must have $n = 1$, which means that s is a local parameter at Q, and $e = p$, which means that P is totally ramified. By (1.1.24) we see that $\mathcal{O}_Q = \mathcal{O}_P[s]$, and since $s \in k' \otimes_k \mathcal{O}_P$, we conclude that P is nonsingular with respect to k'. □

Unfortunately, the possibility that $d(Q|P)$ can be negative is a real one, as shown by Exercise 3.11. When this happens for a scalar extension K'/K, (3.3.5) shows that $g_{K'} < g_K$. The phenomenon of genus reduction under scalar extension is well known. See e.g. [21]. The rather nondescriptive term "conservative" has been used in the literature to describe function fields whose genus is invariant under scalar extension, but it seems more natural and more descriptive to simply call such fields "nonsingular."

Thus, we will say that a function field K/k is singular with respect to a finite extension k'/k if some prime divisor of K is singular with respect to k'. We will say that K is singular if it is singular with respect to some finite scalar extension, and nonsingular otherwise.

Corollary 3.4.6. *Let K/k be a geometric function field. Then K has at most finitely many singular prime divisors. Moreover, there exists a purely inseparable finite extension k'/k such that $k' \otimes_k K$ is nonsingular. In particular, every prime divisor of K has a splitting field that is a finite extension of k.*

Proof. Let $P \in \mathbb{P}_K$ and suppose that K' is a scalar extension of K. If $d(Q|P) < 0$ for some $Q \in \mathbb{P}_{K'}$, then Q divides $\mathscr{D}_{K'/K}$. Moreover, $d(Q'|P) < 0$ for any divisor Q' of Q in any larger scalar extension by (3.3.4) and (3.4.5). Since $g_{K'} \geq 0$, the Riemann-Hurwitz formula (3.3.5) implies that $\deg \mathscr{D}_{K'/K} \geq -2g_K$ and thus K has at most $2g_K$ singular prime divisors.

Furthermore, an obvious induction argument on g_K shows that there is a finite extension k'/k such that $k' \otimes_k K$ is nonsingular. Enlarging k' if necessary, we may assume that k'/k is normal. Let k_0 be the fixed field of $\mathrm{Gal}(k'/k)$. Then k_0/k is purely inseparable and k/k_0 is separable. Since all prime divisors of $K_0 := k_0 \otimes_k K$ are nonsingular with respect to k' by (3.2.3), it follows from (3.4.1) that K_0 is nonsingular.

For any $P \in \mathbb{P}_K$, every prime divisor Q of P in $k' \otimes_k K$ is nonsingular and thus has F_Q as a splitting field by (3.2.4). Then any finite extension of k containing F_Q for each such Q is a splitting field for P. □

Armed with (3.4.6), we can at last show that for a weakly separable extension of a nonsingular function field, the different is nonnegative.

Theorem 3.4.7. *Let K/k be a nonsingular geometric function field and suppose that K'/k' is a finite weakly separable extension. Then $\mathscr{D}_{K'/K} \geq 0$. If $Q \in \mathbb{P}_{K'}$ is singular, then $d(Q|Q \cap K) \neq 0$.*

Proof. If K' is a scalar extension of K, then K' is nonsingular by definition, and hence $\mathscr{D}_{K'/K} = 0$. Thus, if we can prove the theorem when $k = k'$, it will follow in general by (3.3.4), so we may as well assume that $k' = k$.

Let $Q \in \mathbb{P}_{K'}$ be a divisor of $P \in \mathbb{P}_K$ and let k_1/k be finite. Put $K_1' := k_1 \otimes_k K'$ and $K_1 := k_1 \otimes_k K$. Let $Q_1 \in \mathbb{P}_{K_1'}$ be a divisor of Q, and put $P_1 := Q_1 \cap K_1$. From (3.3.4) we have

$$(*) \qquad d(Q_1|Q) + e(Q_1|Q)d(Q|P) = d(Q_1|P) = d(Q_1|P_1) + e(Q_1|P_1)d(P_1|P).$$

Since K is nonsingular, $d(P_1|P) = 0$.

Suppose first that k_1 is a splitting field for Q. Then Q_1 is a point. Moreover, K_1' and K_1 have the same constant field, so P_1 is also a point and thus $d(Q_1|P_1) \geq 0$ by (3.3.6). Since $d(Q_1|Q) \leq 0$ by (3.4.5), we have

$$d(Q|P) = \frac{d(Q_1|P_1) - d(Q_1|Q)}{e(Q_1|Q)} \geq 0,$$

proving the first statement.

Now suppose that $d(Q|P) = 0$ and let k_1 be arbitrary. We need to show that $d(Q_1|Q) = 0$. Since K_1 is a scalar extension of K, it is nonsingular and we have $d(Q_1|P_1) \geq 0$ by the first part of the proof. Since $d(P_1|P) = 0$, $(*)$ yields $d(Q_1|Q) \geq 0$ and thus $d(Q_1|Q) = 0$ by (3.4.5). □

Note that $k(x)$ is always nonsingular, so a consequence of (3.4.6) is that all singular primes of K divide $\mathscr{D}_{K/k(x)}$ for any separating variable $x \in K$. From (3.3.2) we see that $[dx] = \mathscr{D}_{K/k(x)} - 2[x]_\infty$, and therefore we have

Corollary 3.4.8. *If the geometric function field K/k has a separating variable x such that $[dx]$ and $[x]_\infty$ are nonsingular, then K is nonsingular.* □

3.5 Galois Extensions

For any k-embedding $\sigma : K \to K$ of K into itself, it is clear that $v_Q \circ \sigma$ is another discrete k-valuation of K with valuation ring $\sigma^{-1}(\mathscr{O}_Q)$ and maximal ideal $Q^\sigma := \sigma^{-1}(Q)$. There is, however, a potential notational problem here. Consider, for example, the case that k is a finite field of order q and $\sigma(x) = x^q$ for $x \in K$. Here $v_Q \circ \sigma = q v_Q$, while $Q^\sigma = Q$ for all $Q \in \mathbb{P}_K$. This example will be studied in detail in Chapter 5.

In this section, we assume that K'/k' is a Galois extension of K/k. Since the scalar extension $k'K/K$ is clearly normal, $\mathrm{Gal}(K'/K)$ has a normal subgroup $\mathrm{Gal}(K'/k'K)$ which is called the *geometric Galois group*, while $\mathrm{Gal}(K'/K)$ is sometimes called the *arithmetic Galois group*. In any case, $\mathrm{Gal}(K'/K)$ permutes the prime divisors of K' via the action $Q \to Q^\sigma$ given above.

Suppose that Q is a prime of K' and $P := Q \cap K$. Then $\sigma^{-1}(Q) \cap K = P$, so $\mathrm{Gal}(K'/K)$ in fact permutes the set of prime divisors of any prime of K. Our first important fact is that this action is transitive.

Theorem 3.5.1. *Let Q_1 and Q_2 be prime divisors of K' with $P := Q_1 \cap K = Q_2 \cap K$. Then there exists $\sigma \in \mathrm{Gal}(K'/K)$ with $Q_1^\sigma = Q_2$.*

Proof. By the weak approximation theorem there exists an element $x \in K'$ such that $v_{Q_1}(x) = 1$ while $v_Q(x) = 0$ for every other prime Q dividing P. Then for every prime divisor Q of P and every $\sigma \in \mathrm{Gal}(K'/K)$, we have

$$(*) \qquad v_Q(\sigma(x)) = \begin{cases} 1 & \text{if } Q^\sigma = Q_1, \\ 0 & \text{otherwise.} \end{cases}$$

Let $y := \prod_{\sigma \in \mathrm{Gal}(K'/K)} \sigma(x)$. Then

$$v_Q(y) = \sum_{\sigma \in \mathrm{Gal}(K'/K)} v_Q(\sigma(x))$$

for all prime divisors Q of P. Now $(*)$ implies that $v_Q(y) > 0$ if and only if $Q^\sigma = Q_1$ for some $\sigma \in \mathrm{Gal}(K'/K)$. But $y \in Q_1 \cap K = P$, and therefore $v_Q(y) > 0$ for any prime divisor Q of P. □

For the remainder of this section we will use the following notation and assumptions:

Hypothesis 3.5.2. *K' is a Galois extension of K with $G := \mathrm{Gal}(K'/K)$. Q is a prime of K' with $P := Q \cap K$. For $i = 0, 1, \ldots$, we define*

$$\begin{aligned}
e &:= e(Q|P),\\
f &:= f(Q|P,)\\
G_i &:= \{\sigma \in G \mid \sigma(x) \equiv x \mod Q^i \quad \textit{for all } x \in \mathscr{O}_Q\},\\
K_i &:= \{x \in K' \mid \sigma(x) = x \quad \textit{for all } \sigma \in G_i\},\\
Q_i &:= Q \cap K_i.
\end{aligned}$$

F_P (resp. F_Q) is the residue field of P (resp. Q), and F_Q/F_P is separable.[3]

Of course, the hypothesis of separability of the residue field extension is automatic when k is perfect. Note that by (A.0.16) $G_i = \mathrm{Gal}(K'/K_i)$ for $i \geq 0$. Furthermore, observe that G_0 is just the setwise stabilizer of Q in G, and that each G_i is a normal subgroup of G_0 because it is the kernel of the induced map $G_0 \to \mathrm{Aut}(\mathscr{O}_Q/Q^i)$. We call G_0 (resp. K_0) the *decomposition group* (resp. *field*) of Q and G_1 (resp. K_1) the *inertia group* (resp. *field*). For $i > 1$, the G_i are called the *(higher) ramification groups*. It is clear from the definition that $G_i \supseteq G_{i+1}$ for all i.

Given any $\sigma \in G_1$, choose some $x \in \mathscr{O}_Q$ with $\sigma(x) \neq x$ and put $j = v_Q(\sigma(x) - x)$, then $\sigma \notin G_k$ for $k > j$. Since G_1 is a finite group, we see that $G_m = 1$ for some integer m. We will study the decomposition group by analyzing the "layers" G_i/G_{i+1} separately.

Corollary 3.5.3. *Assume* (3.5.2), *then $|G : G_0|$ is the number of distinct prime divisors of P in K'. For any prime divisor Q' of P in K' we have $e(Q'|P) = e$ and $f(Q'|P) = f$. Moreover, $|G_0| = ef$, $e(Q_0|P) = 1 = f(Q_0|P)$, $e(Q|Q_0) = e$, and $f(Q|Q_0) = f$.*

Proof. Let $\{Q_1, \ldots, Q_r\}$ be the set of all distinct prime divisors of P in K'. As a consequence of (3.5.1), there is a bijection between cosets $G_0\sigma$ of G_0 in G and prime divisors Q^σ of P, whence $|G : G_0| = r$. Moreover, it is immediate that

[3] Many authors denote our G_i by G_{i-1}.

$e(Q_i|P) = e$, and $f(Q_i|P) = f$ for all i. Now (2.1.17) implies that $|K' : K| = efr$. Since $|G| = |K' : K|$, we have $|G_0| = ef$.

From (A.0.16) we know that $G_0 = \mathrm{Gal}(K'/K_0)$. By the previous paragraph applied to the extension K'/K_0, we have $e(Q|Q_0)f(Q|Q_0) = |G_0| = ef$, but since $e = e(Q|Q_0)e(Q_0|P)$ and $f = f(Q|Q_0)f(Q_0|P)$ by (1.1.25), we conclude that $e(Q_0|P) = 1 = f(Q_0|P)$, $e(Q|Q_0) = e$, and $f(Q|Q_0) = f$. □

As noted above, G_1 is the kernel of the natural map $\eta : G_0 \to \mathrm{Aut}(\mathcal{O}_Q/Q)$. Since G_0 fixes K elementwise, $\mathrm{im}(\eta)$ lies in $\mathrm{Gal}(F_Q/F_P)$. The important facts here are that F_Q/F_P is a splitting field and that $\mathrm{im}(\eta) = \mathrm{Gal}(F_Q/F_P)$.

Theorem 3.5.4. *Assume* (3.5.2), *then F_Q/F_P is Galois, and the natural map $G_0/G_1 \to \mathrm{Gal}(F_Q/F_P)$ is an isomorphism. In particular, $|G_1| = e$.*

Proof. By (3.5.3) we have $F_{Q_0} = F_P$, so it suffices to prove this result in the special case that $G = G_0$ and $K = K_0$. Now, by (3.5.1) Q is the unique prime divisor of P in K', so $\mathcal{O}_Q$ is the integral closure of $\mathcal{O}_P$ in K' by (1.1.8). Since F_Q/F_P is separable, there is an element $u \in \mathcal{O}_Q$ such that $F_Q = F_P[\bar{u}]$, where $x \mapsto \bar{x}$ denotes the residue class map. Let $f(X)$ be the minimum polynomial of u over K. Then $f(X)$ has coefficients in $\mathcal{O}_P$, since u is integral over $\mathcal{O}_P$. Since K'/K is Galois and has a root in K', all roots of $f(X)$ lie in K' and are also integral, so f factors into linear factors over $\mathcal{O}_Q$. Furthermore, G is transitive on the roots of f, so $\eta(G)$ is transitive on the roots of $\bar{f}$. This implies that $\bar{f}$ is a multiple of the minimum polynomial of $\bar{u}$ over F_P, and that F_Q is the splitting field of $\bar{f}$ over F_P. Thus, F_Q/F_P is Galois, and $\mathrm{im}(\eta)$ is transitive on the conjugates of $\bar{u}$. But the only automorphism of F_Q/F_P fixing $\bar{u}$ is the identity, so $\mathrm{im}(\eta) = \mathrm{Gal}(F_Q/F_P)$. In particular, $|G_0/G_1| = |F_Q : F_P| = f$, and hence $|G_1| = e$. □

Corollary 3.5.5. *Assume* (3.5.2), *and suppose $K \subseteq E \subseteq K'$ is an intermediate field. Put $Q_E := Q \cap E$. Then $e(Q_E|P) = 1$ if and only if $E \subseteq K_1$. In particular, K'/K_1 is totally ramified at Q.*

Proof. From (A.0.16) we have $\mathrm{Gal}(K'/E) = G_E \subseteq G$, where G_E is the subgroup of G fixing E elementwise. By definition, $G_E \cap G_1$ is the inertia group of Q over E. By (1.1.25), $e(Q_E|P) = 1$ iff $e(Q|Q_E) = e$, but by (3.5.4), the latter condition is equivalent to $|G_E \cap G_1| = e = |G_1|$, i.e., to $G_E \supseteq G_1$, which is in turn equivalent to $E \subseteq K_1$. For $E = K_1$, we have in particular $e(Q|Q_1) = e = |G_1| = |K' : K_1|$, so K'/K_1 is totally ramified at Q. □

We next turn to a further analysis of the ramification group G_1 via the filtration $G_1 \supseteq G_2 \supseteq \cdots$. We need first

Lemma 3.5.6. *Assume* (3.5.2), *let $\sigma \in G_1$ and let t be a local parameter at Q. Then for every integer $j \geq 1$ we have $\sigma \in G_j$ iff $\sigma(t) \equiv t \mod Q^j$.*

Proof. Suppose $\sigma(t) \equiv t \mod Q^j$ for some $\sigma \in G_1$ and some $j \geq 1$. We only need to show that $\sigma \in G_j$ since the converse is trivial.

Applying (1.1.24) to the extension K'/K_1 which is totally ramified by (3.5.4), we find that t is integral over $\mathcal{O}_{Q_1}$ and that $\mathcal{O}_Q = \mathcal{O}_{Q_1}[t]$. Let $x \in \mathcal{O}_Q$ and write $x = \sum_i a_i t^i$ with $a_i \in \mathcal{O}_{Q_1}$, then

$$\sigma(x) = \sum_i a_i \sigma(t)^i \equiv \sum_i a_i t^i \mod Q^j. \quad \square$$

Now fix a local parameter t at Q. For $\sigma \in G_1$ we can write

$$\sigma(t) = u_\sigma t,$$

where u_σ is a unit.

Theorem 3.5.7. *Assume* (3.5.2). *The map* $\sigma \mapsto u_\sigma := \sigma(t)/t$ *defines a homomorphism* $G_1 \to F_Q^*$ *whose kernel is* G_2. *In particular,* G_1/G_2 *is cyclic.*

Proof. For any $\sigma' \in G_1$ we have

$$u_{\sigma'\sigma} t = \sigma'(\sigma(t)) = \sigma'(u_\sigma)\sigma'(t) = \sigma'(u_\sigma) u_{\sigma'} t \equiv u_\sigma u_{\sigma'} t \pmod{t^2}.$$

It follows that $u_{\sigma\sigma'} \equiv u_\sigma u_{\sigma'} \pmod{t}$, so we have a homomorphism from G_1 to the multiplicative group of $\mathcal{O}_Q/Q$. Since $\sigma(t)/t \equiv 1 \pmod{t}$ iff $\sigma(t) \equiv t \pmod{t^2}$, the kernel of this map is G_2 by (3.5.6). We conclude that G_1/G_2 is isomorphic to a finite subgroup of the multiplicative group of a field. In particular, there are at most n solutions of the equation $x^n = 1$ in G_1/G_2 for any n, so G_1/G_2 is cyclic by the fundamental theorem of abelian groups. $\square$

We finally consider the structure of the groups G_i/G_{i+1} for $i \geq 2$. For some fixed choice of local parameter t and $\sigma \in G_i$ we have $\sigma(t) = t + x_\sigma t^i$ for some $x_\sigma \in \mathcal{O}_Q$.

Theorem 3.5.8. *Assume* (3.5.2). *For* $\sigma \in G_i$ *and* $i \geq 2$, *the map* $\sigma \mapsto x_\sigma := (\sigma(t) - t)t^{-i}$ *defines a homomorphism of* G_i *into the additive subgroup of* F_Q *whose kernel is* G_{i+1}. *In particular,* G_i/G_{i+1} *is trivial for* $i \geq 2$ *if* $\mathrm{char}(k) = 0$ *and is an elementary abelian* p*-group if* $\mathrm{char}(k) = p > 0$.

Proof. For any $\sigma' \in G_i$ we have

$$t + x_{\sigma'\sigma} t^i = \sigma'(t) + \sigma'(x_\sigma)\sigma'(t)^i = t + x_{\sigma'} t^i + (x_\sigma + y t^i)(t + x_{\sigma'} t^i)^i$$

for some $y \in \mathcal{O}_Q$. Since $i \geq 2$, we have

$$(t + x_{\sigma'} t^i)^i \equiv t^i \pmod{t^{i+1}},$$

and it follows that

$$x_{\sigma'\sigma} t^i \equiv x_{\sigma'} t^i + x_\sigma t^i \pmod{t^{i+1}}.$$

Thus, we have a homomorphism from G_i to the additive group of $\mathcal{O}_Q/Q$, as asserted. Since $x_\sigma \equiv 0 \pmod{t}$ iff $\sigma(t) \equiv t \pmod{t^{i+1}}$, the kernel of this map is G_{i+1} by (3.5.6). Since there are no finite subgroups of the additive group of a field of characteristic zero, G_2 is trivial in this case. When $\mathrm{char}(k) = p > 0$, additive subgroups of F_Q are abelian of exponent p. □

The fact that G_2 is trivial in characteristic zero suggests that the higher ramification groups are related to wild ramification, and that is indeed the case.

Theorem 3.5.9. *Assume* (3.5.2), *and let* $d(Q|P)$ *be the different exponent. Then*

$$d(Q|P) = \sum_{i=1}^{\infty}(|G_i| - 1).$$

In particular, $G_2 = 1$ *if and only if* $Q|P$ *is tamely ramified.*

Proof. Note that the sum is finite, since $G_i = 1$ for almost all i. By (3.5.5) we have $e(Q_1|P) = 1$ and therefore $d(Q_1|P) = 0$ by (3.3.6). Now (3.3.3) implies that $d(Q|P) = d(Q|Q_1)$.

Let t be a local parameter at Q and let $f(X)$ be the minimum polynomial of t over K_1. Then $f(X)$ factors into linear factors over K', and we have

$$f(X) = \prod_{\sigma \in G_1} (X - \sigma(t)),$$

$$f'(X) = \sum_{\tau} \prod_{\sigma \neq \tau} (X - \sigma(t)),$$

whence

$$f'(t) = \prod_{\sigma \neq 1} (t - \sigma(t)).$$

Using (3.3.8), we get

$$d(Q|Q_1) = v_Q(f'(t)) = \sum_{\sigma \neq 1} v_Q(\sigma(t) - t).$$

However, for $\sigma \in G_i \setminus G_{i+1}$, (3.5.6) yields $v_Q(\sigma(t) - t) = i$, so that

$$d(Q|Q_1) = \sum_{i=1}^{\infty} i(|G_i| - |G_{i+1}|) = \sum_{i=1}^{\infty}(|G_i| - 1). \quad \square$$

3.6 Hyperelliptic Functions

As an application of the preceeding results, we consider the case of a function field K/k that is a separable extension of $k(x)$ of degree 2. Such a function field is called *hyperelliptic*, although some authors restrict this term to exclude elliptic functions. Unless specifically stated otherwise, the results of this section apply to the elliptic case as well.

Suppose, then that $x \in K$ is a separating variable and that $|K : k(x)| = 2$. If $y \in K \setminus k(x)$, then y satisifes a quadratic equation $y^2 + f(x)y + g(x) = 0$ over

$k(x)$. If $\mathrm{char}(k) \neq 2$, we can eliminate the linear term by completing the square: Put $y' := y + f(x)/2$, then $y'^2 + g(x) - f(x)^2/4 = 0$. Now multiplying y' by an appropriate square in $k(x)$ and changing notation, we get $y^2 = f(x)$ where $f(x)$ is a square-free polynomial. The situation is different in characteristic 2, so we will consider that case separately below.

Summarizing, we have

Lemma 3.6.1. *Suppose that* $\mathrm{char}(k) \neq 2$ *and* $|K : k(x)| = 2$. *Then* $K = k(x,y)$ *where* $y^2 = f(x)$ *for some square-free polynomial* $f(x)$. □

We will next compute the different $\mathscr{D}_{K/k(x)}$ in the case $\mathrm{char}(k) \neq 2$. We first suppose that $\alpha := a(x) + b(x)y$ is a constant (i.e. algebraic over k) for some $a, b \in k(x)$. Then $T_{K/k(x)}(\alpha) = 2a(x)$, so $a(x) = a_0$ for some $a_0 \in k$, and $N_{K/k(x)}(\alpha) = a_0^2 - b(x)^2 f(x) = a_1$ for some $a_1 \in k$. If $f(x)$ is nonconstant, which we will henceforth assume, it follows that $b(x) = 0$, and we have shown that k is the full field of constants of K.

Since $f(x)$ is square-free, we have $f(x) =: \prod_{i=1}^r p_i(x)$ where the $p_i(x)$ are distinct irreducible polynomials. Let P be a prime of K dividing the prime p of $k(x)$.

Suppose first that $p \neq p_i$ for all i and $p \neq \infty$. Then $f(x)$ is a unit in $\mathscr{O}_P$. Applying v_P to the equation $y^2 = f(x)$, we see that $v_P(y) = 0$. Since $\mathrm{char}(k) \neq 2$, the equation has distinct roots modulo p. Thus, the hypotheses of (1.1.23) apply, and we conclude that $e(P|p) = 1$ and thus $d(P|p) = 0$.

Next, suppose that $p = p_i$. Then $2v_P(y) = e(P|p)v_p(f(x)) = e(P|p)$. This implies that $v_P(y) = 1$ and $e(P|p) = 2$. By (2.1.3), P is the unique prime of K dividing p, $f(P|p) = 1$, and therefore $\deg P = \deg p$.

Finally, if $p = \infty$, then $2v_P(y) = -e(P|\infty)\deg f$. If f has odd degree, this implies that $v_P(y) = -\deg f$, $e(P|\infty) = 2$, P is unique, $f(P|\infty) = 1$, and $\deg P = \deg \infty = 1$. If f has degree $2n$, we can replace x by $x_1 := x^{-1}$ and y by $y_1 := y/x^n$. Then $y_1^2 = f_1(x_1)$ where the reciprocal polynomial f_1 is not divisible by x_1, so the hypotheses of (1.1.23) hold at $x_1 = 0$. It follows that $e(P|\infty) = 1$, and there are two distinct divisors of ∞ each of degree one, or one divisor of degree two, depending on whether or not the leading coefficient of $f(x)$ is a square in k or not. We have proved

Theorem 3.6.2. *Suppose that* $K = k(x,y)$, $\mathrm{char}(K) \neq 2$, *and*

$$y^2 = f(x) = \prod_{i=1}^{r} p_i(x),$$

where the p_i *are distinct irreducible polynomials and* $r \geq 1$. *Then* k *is the full field of constants of* K, *and for each* i *there is a unique prime* P_i *of* K *dividing* p_i *with* $e(P_i|p_i) = 2$, $f(P_i|p_i) = 1$, *and* $\deg P_i = \deg p_i$, *and we have*

$$[y]_0 = \sum_{i=1}^{r} P_i.$$

If $\deg f$ *is even, we have*

$$\mathscr{D}_{K/k(x)} = \sum_{i=1}^{r} P_i,$$

and $P_\infty := N^*_{K/k(x)}(\infty)$ *has degree 2 and is either prime or the sum of two points.*

If $\deg f(x)$ *is odd, there is a unique prime* P_∞ *dividing* ∞ *with* $e(P_\infty|\infty) = 2$, $f(P_\infty|\infty) = 1$, $\deg P_\infty = 1$, *and we have*

$$\mathscr{D}_{K/k(x)} = P_\infty + \sum_{i=1}^{r} P_i.$$

In particular,

$$\deg \mathscr{D}_{K/k(x)} = \begin{cases} \deg f & \text{if } \deg f \text{ is even,} \\ \deg f + 1 & \text{if } \deg f \text{ is odd.} \end{cases} \qquad \square$$

We can now apply the Riemann–Hurwitz formula (3.3.5) to obtain

Corollary 3.6.3. *Assume the hypotheses of* (3.6.2). *Then*

$$g_K = \begin{cases} \frac{1}{2}(\deg f - 2) & \text{if } \deg f \text{ is even,} \\ \frac{1}{2}(\deg f - 1) & \text{if } \deg f \text{ is odd.} \end{cases}$$

Proof. From (3.3.5) we obtain

$$2g_K - 2 = -4 + \deg \mathscr{D}_{K/k(x)},$$

and the result follows from (3.6.2). $\square$

Finally, we obtain a basis for the space of regular differential forms. Continuing the hypotheses of (3.6.2), we can differentiate the defining equation to obtain

$$2ydy = f'(x)dx,$$

whence

$$(*) \qquad [y] + [dy] = [f'(x)] + [dx].$$

From (3.6.2), we know that $[y]_0 = \sum_i P_i$. Moreoever, $\gcd(f(x), f'(x)) = 1$ since $f(x)$ has distinct irreducible factors. This implies that $f(x)$ and $f'(x)$ have disjoint zeros in K, and therefore y and $f'(x)$ also have disjoint zeros in K. Now $(*)$ implies that

$$(*) \qquad [y]_0 \leq [dx]_0.$$

The only poles of x in K are at the prime(s) at infinity. When $\deg f$ is odd, $[x]_\infty = 2P_\infty$, where P_∞ has degree 1, while when $\deg f$ is even, $[x]_\infty = P_\infty$ has degree 2. Looking at the respective Laurent series, we conclude that

$$[dx]_\infty = \begin{cases} 3P_\infty & \text{if } \deg f \text{ is odd,} \\ 2P_\infty & \text{if } \deg f \text{ is even.} \end{cases}$$

From $(*)$ we have $[dx] \geq [y]_0 - [dx]_\infty$, but using (3.6.3) we get

$$\deg[dx] \geq \deg f - \deg[dx]_\infty = 2g_K - 2.$$

Since $[dx]$ is a canonical divisor, we must have equality, which implies that $(*)$ is an equality. In particular, $v_{P_i}(dx/y) \geq 0$ for all i.

The only other possible pole of dx/y is at infinity, but here we actually have a zero of order $\deg f - 3 = 2g - 2$ when $\deg f$ is odd, or of order $\deg f - 4 = 2g - 2$ in the even case. So not only is dx/y regular, in fact $x^i dx/y$ is regular for $0 \leq i < g$. Since x is transcendental, these forms are linearly independent over k, and therefore are k-basis for $\Omega_K(0)$ by (2.5.10). As a consequence, we note that the ratio of any two regular forms lies in $k(x)$, and if $g_K \geq 2$ these ratios generate $k(x)$. We have proved

Theorem 3.6.4. *Suppose $|K : k(x)| = 2$ for some $x \in K$ and $\operatorname{char}(k) \neq 2$. Then $\{x^i dx/y \mid 0 \leq i < g_K\}$ is a k-basis for $\Omega_K(0)$. Furthermore, if $g_K \geq 2$ then $k(x)$ is the subfield of K generated by all ratios ω'/ω of regular differential forms, and is thus the unique rational subfield of K of index 2.* □

We remark that the even-degree case can be completely avoided provided that $p_i = ax - b$ is linear for at least one i. Then the substitution

$$x_1 := \frac{1}{ax - b}$$

will produce an equation $y_1 = f_1(x_1)$, where f_1 is square-free of odd degree.

For the remainder of this section we will assume that k is a perfect field of characteristic two. Beginning with the equation $y^2 + f(x)y + g(x) = 0$, we can replace y by $y/f(x)$ and change notation to obtain $y^2 + y = f(x)$. Note that the other root of this equation is $y + 1$. We see that y is an element of K of trace 1 and norm $f(x)$. Suppose $y_1 = a(x) + b(x)y$ is another element of K of trace 1. Since $T_{K/k(x)}$ is $k(x)$-linear, and is zero on $k(x)$, we get $b(x) = 1$. Then $N_{K/k(x)}(y_1) = a(x)^2 + a(x) + f(x)$. Conversely, the equation $Y^2 + Y = a(x)^2 + a(x) + f(x)$ has roots $y + a(x)$ and $y + a(x) + 1$ in K. In other words, we have proved

Lemma 3.6.5. *Suppose that $\operatorname{char}(k) = 2$ and $K/k(x)$ is separable of degree 2. If $k(x)^{(2)}$ denotes the additive subgroup of $k(x)$ consisting of all rational functions of the form $a(x)^2 + a(x)$, then there is a one-to-one correspondence between cosets $f(x) + k(x)^{(2)}$ in $k(x)$ and separable extensions $K/k(x)$ of degree 2, under which K corresponds to $f(x) + k(x)^{(2)}$ if and only if there exists an element $y \in K$ of trace 1 and norm $f(x)$. In particular, $K = k(x,y)$, where $y^2 + y = f(x)$ for some $f(x) \in k(x)$.* □

We first consider the (uninteresting) case that K is a scalar extension of $k(x)$. Then there is a constant $\alpha \in K \setminus k(x)$. Since $T_{K/k(x)}(\alpha)$ is a nonzero element of k, we can divide to get a scalar of trace 1. Using (3.6.5) we get

Lemma 3.6.6. *Suppose that* $\operatorname{char}(k) = 2$ *and* $K/k(x)$ *is separable of degree* 2. *Then* $K/k(x)$ *is a scalar extension if and only if* $f(x) \equiv c \pmod{k(x)^{(2)}}$ *for some constant* $c \in K$. □

We next compute the different. As before, let P be a prime of K dividing the prime p of $k(x)$. Note that $v_P(y) < 0$ iff $v_P(y+1) < 0$ iff $v_P(y^2+y) < 0$ iff $v_p(f(x)) < 0$. In particular, if $v_p(f(x)) \geq 0$, then $y \in \mathcal{O}_P$. Moreover, y satisfies a polynomial with coefficients in $\mathcal{O}_p$ and distinct roots modulo p, so (1.1.23) implies that $e(P|p) = 1$ and hence $d(P|p) = 0$.

Suppose, then, that $f(x)$ has a pole at p of order n. Then $v_P(y)$ must be negative and hence $v_P(y) = v_P(y+1)$, and we get

$$2v_P(y) = v_P(f(x)) = -e(P|p)n.$$

Suppose that $n = 2d-1$ is odd. Then we can conclude that $e(P|p) = 2$ and $v_P(y) = -n$. Since $|K : k(x)| = 2$, we have $f(P|p) = 1$. Furthermore, if we put $s := p^d y$, we have $v_P(s) = dv_P(p) - n = 1$, so s is a local parameter and

$$s^2 + p^d s = p^{2d}(y^2+y) = p^{2d} f(x) = pu(x)$$

for some p-local unit $u(x)$. Since $K/k(x)$ is totally ramified at p, (3.3.8) implies that $d(P|p) = 2d$.

So far, we have seen that when $f(x)$ is a p-local integer, p is unramified and hence $d(P|p) = 0$, while if $f(x)$ has a pole of odd order $2d-1$ at p, p is ramified and $d(P|p) = 2d$. Fortunately, it turns out that we don't have to consider poles of even order at all, as we now argue.

We first recall the so-called partial fractions algorithm: Given a rational function whose denonimator is a product $h_1(x)h_2(x)$ of relatively prime polynomials, we can find polynomials $a_1(x), a_2(x)$ with $a_1(x)h_1(x) + a_2(x)h_2(x) = 1$ and write

$$\frac{1}{h_1(x)h_2(x)} = \frac{a_1(x)}{h_2(x)} + \frac{a_2(x)}{h_1(x)}.$$

By repeated applications of the Euclidean algorithm, therefore, we can write any rational function as a sum

$$(*) \qquad f(x) = \sum_{i=1}^{m} \frac{a_i(x)}{p_i(x)^{2d_i-1}} + \sum_{i=m+1}^{n} \frac{a_i(x)}{p_i(x)^{2d_i}},$$

where the a_i are polynomials, the d_i are positive integers, and the p_i are distinct irreducible polynomials. If $m = n$, we are happy. If $m < n$, we need

Lemma 3.6.7. *Suppose that k is perfect of characteristic* 2, $a(x), p(x) \in k[x]$, $p(x)$ *is irreducible, and*

$$f(x) := \frac{a(x)}{p(x)^{2e}}$$

for some positive integer e. Then there exist polynomials $b(x), c(x), d(x)$, *and* $p_1(x)$ *with* $\deg p_1(x)^2 < \deg p(x)$ *such that*

$$f(x) \equiv \frac{b(x)}{p(x)^e} + \frac{c(x)}{p(x)^{2e-1}} + \frac{d(x)}{p_1(x)^2} \pmod{k(x)^{(2)}}.$$

Proof. Separating terms of even and odd degree and using that fact that $k^2 = k$, we can write $a(x) = a_0(x)^2 + xa_1(x)^2$ and $p(x) = p_0(x)^2 + xp_1(x)^2$ where a_i and p_i are polynomials ($i = 1,2$). Note that $\deg p_1^2 < \deg p$ and hence that $p_1(x)$ and $p(x)$ are relatively prime. Then

$$x = \frac{p_0^2}{p_1^2} + \frac{p}{p_1^2},$$

and

$$\frac{a}{p^{2e}} = \frac{a_0^2}{p^{2e}} + \frac{a_1^2 p_0^2}{p^{2e} p_1^2} + \frac{a_1^2}{p^{2e-1} p_1^2} \equiv \frac{a_0}{p^e} + \frac{a_1 p_0}{p^e p_1} + \frac{a_1^2}{p^{2e-1} p_1^2} \pmod{k(x)^{(2)}}.$$

Using partial fractions and collecting terms, the result follows. □

Suppose $m < n$ in (*) and there are k terms in the sum whose square denominator is of highest degree $2e$. Applying the lemma to one of these terms and rewriting the resulting expression in the same form, it is clear that we now have $k-1$ terms whose denominators have degree $2e$. After a finite number of steps, we have found an element $y \in K$ with $y^2 + y = a(x) + b(x)/c(x)$ such that $a(x), b(x), c(x)$ are polynomials, every prime divisor of c occurs to an odd power, and $\deg b < \deg c$. Since every polynomial is clearly congruent to a polynomial of odd degree modulo $k(x)^{(2)}$, we have proved

Theorem 3.6.8. *Let k be perfect of characteristic* 2 *and suppose that* $K/k(x)$ *is separable of degree* 2. *Then* $K = k(x,y)$, *where* $y^2 + y = f(x)$ *and all poles of* $f(x)$ *are of odd order. If* $f(x)$ *has a pole of order* $2d_i - 1$ *at* p_i *for* $1 \le i \le n$ *and no other poles, then for each i there is a unique prime divisor* P_i *in K of* p_i *and it satisfies* $e(P_i|p_i) = 2$ *and* $f(P_i|p_i) = 1$. *Moreover, no other primes of* $k(x)$ *are ramified in K, and*

$$\mathscr{D}_{K/k(x)} = \sum_{i=1}^{n} 2d_i P_i. \quad \square$$

Indeed, the above discussion yields a constructive algorithm for finding $f(x)$, and therefore for finding $\mathscr{D}$. Furthermore, the Riemann–Hurwitz formula immediately implies

Corollary 3.6.9. *With the hypotheses and notation of the theorem,*

$$g_K = -1 + \sum_{i=1}^{n} d_i \deg p_i. \quad \square$$

Finally, we would like to find a basis for $\Omega_K(0)$, the space of regular differential forms.

Theorem 3.6.10. *Let k be perfect of characteristic 2, and let $K := k(x,y)$, where $y^2+y=f(x)$ and the pole divisor of $f(x)$ in $k(x)$ is given by*

$$[f(x)]_\infty = \sum_{i=1}^{n}(2d_i-1)p_i.$$

Define

$$\phi(x) := \prod_{p_i \neq \infty} p_i(x)^{d_i},$$

where the product runs over the finite poles of $f(x)$ only, and put

$$\omega = \frac{dx}{\phi(x)}.$$

Then $\{\omega, x\omega, \ldots, x^{g-1}\omega\}$ is a basis for the space of regular differential forms. In particular, if $g_K \geq 2$, $k(x)$ is the unique rational subfield of K of index 2.

Proof. The zero divisor of $\phi(x)$ in K is

$$[\phi(x)]_0 = \sum_{P_i \neq \infty} 2d_iP_i = \mathscr{D} - 2d_\infty P_\infty,$$

where we set $d_\infty = 0$ if $f(x)$ has no pole at infinity. Thus, $\phi(x)$ has a pole of order $\deg \mathscr{D} - 2d_\infty$ at infinity.

Since dx has no zeros in $k(x)$ and a pole of order 2 at infinity, it follows from (3.3.2) that

$$[dx] = A + \mathscr{D},$$

where $A = -2N^*(p_\infty)$ has degree -4. Thus, dx has a pole at infinity of order $4 - 2d_\infty$. We conclude that the form

$$\omega := \frac{dx}{\phi(x)}$$

is regular and has a zero at infinity of order $\deg \mathscr{D} - 4 = 2g_K - 2$. Moreover, in the case that there are two points of K at infinity, each point has equal multiplicity $g_K - 1$. Since the pole divisor of x in K has degree two in all cases, and is a sum of two simple poles when there are two points at infinity, it follows that $x^i\omega$ is regular for $0 \leq i < g_K$. □

3.7 Exercises

Exercise 3.1. Let K'/K be a finite, separable extension of function fields and let $P \in \mathbb{P}_K$. Let $M \subseteq K'$ be a finitely generated $\mathscr{O}_P$-module of rank $n = |K' : K|$, and define

$$M^* = \{x \in K' \mid \mathrm{tr}_{K'/K}(xm) \in \mathscr{O}_P \text{ for all } m \in M\}.$$

(i) Let $\{m_1,\ldots,m_n\}$ be an $\mathcal{O}_P$-basis for M and define

$$\Delta_M := v_P(\det(\mathrm{tr}_{K'/K}(m_i m_j))).$$

Show that Δ_M is independent of the choice of basis for M.

(ii) Suppose that $M \subseteq M' \subseteq M^*$, where M' is also finitely generated. Let $\ell(M'/M)$ be the length of M'/M (see (1.1.13)). Prove that

$$\Delta_M = \Delta_{M'} + 2\ell(M'/M).$$

(iii) Conclude that M^* is a finitely generated $\mathcal{O}_P$-module and that M^*/M has length at most $\Delta_M/2$.

Exercise 3.2. Use Exercise 3.1 to give an alternative proof of (2.1.18) in the separable case.

Exercise 3.3. Let K/k be a function field and suppose that k' is a finite, separable extension of k. Let $P \in \mathbb{P}_K$ and put $K' := k' \otimes_k K$. Use Exercise 3.1 to show that the integral closure of $\mathcal{O}_P$ in K' is just $k'\mathcal{O}_P$.

Exercise 3.4. Generalize Exercises 3.1 and 3.3 by replacing $\mathcal{O}_P$ by the intersection of a finite number of valuation rings of K.

Exercise 3.5. Let K'/k' be a finite separable extension of K/k with $Q \in \mathbb{P}_{K'}$ and $P := Q \cap K$, and assume that F_Q/F_P is separable. The object of this exercise is to prove that (3.3.11) holds under this weaker hypothesis.

(i) Show that there exists a finite separable extension k_1/k such that if we put $K_1' := k_1 \otimes_{k'} K'$ and $K_1 := k_1 K \subseteq K_1'$, we can choose a prime $Q_1 \in \mathbb{P}_{K_1'}$ with $Q = Q_1 \cap K'$ and $f(Q_1|Q_1 \cap K_1) = 1$.

(ii) Put $P_1 := Q \cap K_1$. Show that $d(Q|P) = d(Q_1|P_1)$.

(iii) Let R', R_1, and R_1' denote the integral closure of $\mathcal{O}_P$ in K', K_1, and K_1', respectively. By replacing $\mathcal{O}_{P_1}$ by a finite intersection of valuation rings of K_1, generalize the argument of (3.3.11) to show that

$$\min_{x \in (R_1')^*} v_{Q_1}(x) = -d(Q_1|P_1).$$

(iv) Use Exercise 3.4 to complete the proof.

Exercise 3.6. Let k be algebraically closed of characteristic $p > 0$ and let $u \in k(t)$ such that $k(t)/k(u)$ is a Galois extension whose Galois group G is a p-group. Prove that $G_0 = G_1 = G_2$ and that $G_3 = 1$.

Exercise 3.7. Let k be algebraically closed of characteristic zero, and let K be a finite extension of $k((t))$ of degree n.

(i) Use Exercise 1.8 to show that the prime (t) is totally ramified in K'.

(ii) Prove that every Galois extension of $k((t))$ is cyclic.

(iii) Prove that $K = k((t^{1/n}))$.

Exercise 3.8. Suppose that K'/k' and K/k are function fields, and $K' = K(y)$ where $y^n = x \in K$ for some integer n relatively prime to $\mathrm{char}(k)$. Such an extension is called a *Kummer extension*. Assume that $x \notin K^d$ for any divisor d of n.

(i) Prove that $|K' : K| = n$.

(ii) Let $Q \in \mathbb{P}_{K'}$ be a divisor of $P \in \mathbb{P}_K$. Put $d_P := \gcd(v_P(x), n)$. Prove that

$$e(Q|P) = \begin{cases} n/d_P & \text{for } d_P > 0, \\ 1 & \text{for } d_P = 0. \end{cases}$$

[Hint: Consider the intermediate field $K(y^d)$.]

(iii) Prove that

$$g_{K'} = 1 + \frac{n}{|k' : k|}\left(g_K - 1 + \frac{1}{2}\sum_{P \in \mathbb{P}_K}\left(1 - \frac{d_P}{n}\right)\deg P\right).$$

Exercise 3.9. We say that an extension K'/K is *unramified* if $e(Q|Q \cap K) = 1$ for every prime $Q \in \mathbb{P}_{K'}$. Suppose that n is relatively prime to $\mathrm{char}(K)$ and that $K' = K(y)$ for some $y \in K'$ with $y^n = x \in K$. Prove that K'/K is unramified if and only if $\lfloor x \rfloor = nD$ for some divisor $D \in \mathrm{Div}(K)$. [Hint: (1.1.23).]

Exercise 3.10. Let K be an elliptic function field over an algebraically closed field k of characteristic 2. Prove the following:

(i) $K = k(x,y)$ where $y^2 + y = f(x)$, and $f(x)$ has either exactly one pole of order 3 or exactly two simple poles.

(ii) By using substitutions of the form $y' = y + g(x)$ and

$$x' = \frac{ax+b}{cx+d}$$

for $a,b,c,d \in k$ with $ad - bc \neq 0$, the first case can be reduced to

$$y^2 + y = x^3,$$

and the second case can be reduced to

$$y^2 + y = x + \frac{\lambda}{x}$$

for some nonzero $\lambda \in k$, uniquely determined by K (compare with Exercise 2.8).

Exercise 3.11. Let k_0 be a field of characteristic $p > 2$ and let $k = k_0(t)$ where t is transcendental over k_0. Let $K := k(x,y)$ where $y^2 = x^p - t$. Show that $g_K = (p-1)/2$, but that $k(t^{1/p}) \otimes_k K$ has genus zero.

Exercise 3.12. Let k_0 be a field of characteristic $p > 0$ and let $k = k_0(s,t)$, where s and t are transcendental over k_0. Let $K := k(x,y)$ where $y^p = x^{p+1} + tx^p + s$.

(i) Show that there is a unique prime divisor $P \in \mathbb{P}_K$ dividing the prime (x) of $k[x]$ and that $F_P \simeq k(s^{1/p})$.

(ii) Let $k' := k(s^{1/p})$ and let $K' := k' \otimes_k K$. Show that there is a unique prime Q of K' dividing P and that $F_Q \simeq k(t^{1/p})$.

(iii) Show that $(y - s^{1/p})/x$ is integral over $\mathcal{O}_P$ but does not lie in $k' \otimes_k \mathcal{O}_P$.

4

Projective Curves

In this chapter we make contact with the classical theory of algebraic curves in projective space. For simplicity we will assume throughout this chapter that the ground field k is algebraically closed. Since we do not have at our disposal the machinery of algebraic geometry, our treatment here is necessarily somewhat ad hoc. The preferred approach to this subject is via the theory of schemes and varieties.

4.1 Projective Varieties

Recall that n-dimensional projective space $\mathbb{P}^n(k)$ is defined as the set of lines through the origin in k^{n+1}. We denote the line through $(a_0, a_1 \dots, a_n)$ by $(a_0 : a_1 : \cdots : a_n)$. Each such line is called a point of $\mathbb{P}^n$. The zero set of a homogeneous polynomial $f \in k[X_0, \dots, X_{n+1}]$ is a union of lines through the origin, and therefore defines a subset $\mathbf{V}(f) \subseteq \mathbb{P}^n$. The set of common zeros of an arbitrary set of homogeneous polynomials in $\mathbb{P}^n$ is called a *closed* set. It is easy to check that the closed sets define a topology on $\mathbb{P}^n$ called the *Zariski topology*.

Conversely, an arbitrary polynomial $f \in k[X_0, \dots, X_n]$ can be written uniquely as a sum of homogeneous polynomials of distinct degrees:

$$f = \sum_{d=0}^{m} f_d.$$

If $a := (a_0, a_1, \dots, a_n) \in k^{n+1}$ and f vanishes on the line $\{\lambda a \mid \lambda \in k\}$, we have

$$0 = f(\lambda a) = \sum_{d=0}^{m} \lambda^d f_d(a)$$

for all $\lambda \in k$, from which we conclude that $f_d(a) = 0$ for all d. It follows that the ideal $\mathbf{I}(V)$ of all polynomials in $k[X_0, \ldots, X_n]$ which vanish at an arbitrary subset $V \subseteq \mathbb{P}^n$ is a *graded ideal*, i.e. it is the direct sum

$$\mathbf{I}(V) = \bigoplus_{d=0}^{\infty} \mathbf{I}(V)_d$$

of homogeneous subspaces. In dealing with closed subsets of projective space, we are thus naturally led to the categories of graded k-algebras and modules. Briefly, recall that a k-vector space V is *graded* if we are given a direct sum decomposition

$$V = \bigoplus_{d \in \mathbb{Z}} V_d,$$

although it is frequently the case that $V_d = 0$ for $d < 0$. The elements of V_d are called homogeneous of degree d. A homogeneous map $\phi : V \to W$ of graded vector spaces (of degree d) is a map of k-algebras such that for some integer d we have $\phi(V_e) \subseteq W_{d+e}$ for all $e \in \mathbb{Z}$. A *graded k-algebra* is a k-algebra A that is a graded k-vector space such that $A_d A_e \subseteq A_{d+e}$ for all $d, e \in \mathbb{Z}$. In particular, $k \subseteq A_0$. A map of graded k-algebras is a k-algebra homomorphism that is homogeneous of degree zero.

If A is a graded k-algebra, a *graded A-module M* is an A-module that is a graded k-vector space such that $A_d M_e \subseteq M_{d+e}$ for all $d, e \in \mathbb{Z}$. In particular, the regular module is a graded module. A graded submodule $N \subseteq M$ is an A-submodule with a grading such that the inclusion map is homogeneous of degree zero.

It is easy to check that the kernel and image of a homogeneous map are graded submodules, and that the quotient of a graded module by a graded submodule has a natural grading such that the quotient map is homogeneous of degree zero. In particular, the quotient of a graded k-algebra by a graded ideal is a graded k-algebra. It is also straightforward to verify

Lemma 4.1.1. *Suppose A is a graded k-algebra, M is a graded A-module, and $N \subseteq M$ is an arbitrary A-submodule. Then $N = \bigoplus_d N \cap M_d$ if and only if N is generated by homogeneous elements.*

Since the polynomial ring $A := k[X_0, \ldots, X_n]$ is a graded k-algebra, any graded k-algebra which is generated by $n+1$ elements of degree one is a graded quotient of A.

Returning now to the Zariski topology, we say that a closed set is *irreducible* if it cannot be written as a union of proper closed subsets. For example, if the homogeneous polynomial f is a product $f = f_1 f_2$ of two distinct irreducible polynomials (both necessarily homogeneous), it is clear that $\mathbf{V}(f) = \mathbf{V}(f_1) \cup \mathbf{V}(f_2)$. But is $\mathbf{V}(f_1)$ proper? For this and other important facts about the zeros of multivariate polynomials we need

Theorem 4.1.2. *If the finitely generated k-algebra K is a field, then $|K : k|$ is finite.*

Proof. Let $K = k[x_1, x_2, \ldots, x_n]$ and put $k_1 := k(x_1) \subseteq K$. Then $K = k_1[x_2, \ldots, x_n]$, and $|K : k_1|$ is finite by induction on n. If x_1 is algebraic over k, then $|k_1 : k|$ is finite and we are done.

Suppose then, by way of contradiction, that x_1 is transcendental. Then k_1 is isomorphic to the field of rational functions in one variable. For $x \in K$ let $M(x) = M_{K/k_1}(x)$ be the matrix of the k_1-linear transformation $y \mapsto xy$ with respect to some basis for K/k_1. Then the map $x \mapsto M(x)$ defines a k-algebra monomorphism from K into the ring of $m \times m$ matrices over k_1, and $M(x) = xI$ for $x \in k_1$.

Now, the polynomial ring in one variable has infinitely many irreducibles over any field, so we can choose an irreducible polynomial $p(x_1) \in k[x_1]$ such that p does not divide the denominator of any entry of $M(x_i)$ for all i. But this implies that p does not divide the denominator of any entry of $M(x)$ for any $x \in K$, because the matrices $M(x_i)$ generate $M(K)$ as a k-algebra. Hence, $p^{-1}I \neq M(x)$ for any $x \in K$, a contradiction that completes the proof. □

Corollary 4.1.3. *Suppose that M is a maximal ideal of the polynomial ring $A = k[X_1, \ldots, X_n]$. Then there exist $a_1, \ldots, a_n \in k$ such that $M = (X_1 - a_1, X_2 - a_2, \ldots, X_n - a_n)$. A polynomial f lies in M iff $f(a_1, a_2, \ldots, a_n) = 0$. A necessary and sufficient condition for a set of polynomials to generate a proper ideal of A is that they all have a common zero.*

Proof. A/M satisfies the hypotheses of (4.1.2). Since k is algebraically closed, we conclude that $A/M = k$. Let a_i be the image of X_i in A/M. Then the natural map $A \to A/M = k$ is just evaluation at the point $(a_1, \ldots, a_n)$. □

Given an ideal $J \subseteq k[X_0, \ldots, X_n]$, let $\mathbf{V}(J) \subseteq k^{n+1}$ denote the zero set of J, and for a subset $S \subseteq k^{n+1}$ write $\mathbf{I}(S)$ for the ideal of all functions that vanish at S. Note that $\mathbf{V}(\mathbf{I}(S))$ is just the closure of S in the Zariski topology. In general, the ideal $\mathbf{I}(\mathbf{V}(J))$ is larger than J. For example, it contains

$$\sqrt{J} := \{f \in A \mid f^r \in J \text{ for some positive integer } r\}.$$

In fact, we have

Corollary 4.1.4 (Hilbert Nullstellensatz). *If J is an ideal of $A := k[X_0, X_1, \ldots, X_n]$, then $\mathbf{I}(\mathbf{V}(J)) = \sqrt{J}$.*

Proof. If $f^r \in J$, it is clear that f vanishes at every zero of J. The converse argument is a well-known but clever trick. We adjoin an additional indeterminate Y to A. Then the ideal $J + (1 - fY) \subseteq A[Y]$ has no zeros at all; hence $A[Y] = J + (1 - fY)$ by (4.1.3). Let $\overline{A} = A/J$ and let $\bar{f} := f + J \in \overline{A}$. Then it follows that $1 - \bar{f}Y$ is invertible in $\overline{A}[Y]$. Thus, there exists an integer r and elements $a_0, \ldots, a_{r-1} \in \overline{A}$ such that $(1 - \bar{f}Y)(a_0 + a_1Y + \cdots + a_{r-1}Y^{r-1}) = 1$. Equating coefficients of Y^i, we find that $a_0 = 1$, $a_i = \bar{f}^i$ for $1 \leq i < r$ and finally that $\bar{f}^r = 0$. □

We say that an ideal $J \subseteq A$ is a *radical ideal* if $\sqrt{J} = J$. Note that J is a radical ideal if and only if A/J has no nilpotent elements.

We want to apply the Nullstellensatz to graded ideals and closed subsets of projective space. Note, however, that while every proper ideal of $k[X_0, X_1, \ldots, X_n]$ has a nontrivial zero in k^{n+1}, the unique maximal graded ideal consisting of all polynomials with zero constant term vanishes only at the origin and therefore has no proper zeros in $\mathbb{P}^n$. By a *proper* graded ideal, therefore, we will mean a nonzero graded ideal whose radical is not the unique maximal graded ideal. By abuse of notation, we will write $\mathbf{V}(J) \subseteq \mathbb{P}^n$ when J is a proper graded ideal. If J is graded and proper, then, as we have observed above, $\mathbf{V}(J)$ is a union of lines and thus $\mathbf{I}(\mathbf{V}(J))$ is also graded and proper. So we have

Corollary 4.1.5. *The radical of a proper graded ideal is a proper graded ideal.* □

The main consequence of the Nullstellensatz for our purposes is

Corollary 4.1.6. *The mappings $I \mapsto \mathbf{V}(I)$ and $V \mapsto \mathbf{I}(V)$ define an inclusion-reversing bijection between proper graded radical ideals $I \subseteq k[X_0, \ldots, X_n]$ and closed subsets $V \subseteq \mathbb{P}^n$. In particular, a closed set is the union of two proper closed sets iff the corresponding radical ideal is the intersection of two properly larger graded radical ideals.*

Proof. Because the two mappings are inclusion-reversing, we have $\mathbf{V}(\mathbf{I}(\mathbf{V}(J))) = \mathbf{V}(J)$ and $\mathbf{I}(\mathbf{V}(\mathbf{I}(S))) = \mathbf{I}(S)$ for any ideal $J \subseteq k[X_0, \ldots, X_n]$ and any subset $S \subseteq \mathbb{P}^n$. Thus $\mathbf{I}$ and $\mathbf{V}$ are bijective mappings between all closed subsets of $\mathbb{P}^n$ and all ideals of the form $\mathbf{I}(S)$. The ideal $\mathbf{I}(S)$ is certainly a radical ideal, and by (4.1.4), if J is any radical ideal then $J = \sqrt{J} = \mathbf{I}(\mathbf{V}(J))$, so the ideals of the form $\mathbf{I}(S)$ are precisely the radical ideals. □

We call a closed set in $\mathbb{P}^n$ a *projective variety* if it is irreducible, that is, if it is not the union of two proper closed subsets.

Lemma 4.1.7. *A closed set V is a variety iff $\mathbf{I}(V)$ is a prime ideal.*

Proof. It is clear that a prime ideal cannot be the intersection of two properly larger ideals. Conversely, suppose $V := \mathbf{V}(I) \subseteq \mathbb{P}^n$ is irreducible, and $I := \mathbf{I}(V)$. Put $A := k[X_0, \ldots, X_n]$, choose homogeneous elements $x_1, x_2 \in A \setminus I$, and put $V_i = \mathbf{V}(I, x_i)$. Then V_1 and V_2 are proper closed subsets of the irreducible closed set V, so there must be an element $v \in V \setminus V_1 \cup V_2$. Then $x_1(v)x_2(v) \neq 0$, so $x_1x_2 \notin I$. We have shown that in the graded algebra A/I, the product of nonzero homogeneous elements is nonzero. If y_1, y_2 are any two nonzero elements of A/I, and x_1, x_2 are their lowest-degree homogeneous components, then x_1x_2 is the lowest-degree homogeneous component of y_1y_2 and thus $y_1y_2 \neq 0$. □

Suppose the graded k-algebra A is an integral domain. We define the *homogeneous field of fractions* of A to be the subset of the full field of fractions consisting of all fractions x/y where x and y are homogeneous of the same degree. It is easy to check that this is in fact a subfield of the full field of fractions.

Suppose now that $V \subseteq \mathbb{P}^n$ is a projective variety. While it doesn't make sense to evaluate polynomials at points of $\mathbb{P}^n$, we can evaluate a quotient f/g of two homogeneous polynomials of the same degree at a point of $\mathbb{P}^n$. In this way, we get a k-valued function defined at each point $(a_0 : \cdots : a_n)$ of $\mathbb{P}^n$ at which $g(a_0,\dots,a_n) \neq 0$. Restricting to V, we get a function $V \setminus \mathbf{V}(g) \to k$ that depends only on the cosets $f+\mathbf{I}(V)$ and $g+\mathbf{I}(V)$. Speaking somewhat loosely, we will say that f/g restricts to a rational function on V.

Since $\mathbf{I}(V)$ is a prime ideal, the graded k-algebra

$$k[V] := k[X_0,\dots,X_n]/\mathbf{I}(V)$$

is an integral domain, and we let $k(V)$ be the homogeneous field of fractions of $k[V]$. Each element of $k(V)$ defines a rational function on V. For this reason, $k(V)$ is called the field of rational functions on V.

Lemma 4.1.8. *Let $V \subseteq \mathbb{P}^n$ be a projective variety. Relabeling the X_i if necessary so that $V \nsubseteq \mathbf{V}(X_0)$, $k(V)$ is generated as a field over k by the restrictions of the functions X_i/X_0 to V for $1 \leq i \leq n$.*

Proof. Abusing notation, we continue to denote by X_i/X_0 the restriction of X_i/X_0 to V. Put $K_0 := k(X_1/X_0,\dots,X_n/X_0)$. Then clearly $K_0 \subseteq k(V)$. However, if $p(X)$ is homogeneous of degree d, we have $p(X)/X_0^d \in K_0$ and therefore $p(X)/q(X) \in K_0$ for every homogeneous polynomial $q(X)$ of degree d that does not vanish at V. $\square$

Let $k[V]_0$ denote the k-subalgebra of $k(V)$ generated by the restrictions of the functions X_i/X_0 to V for $1 \leq i \leq n$. The subset $V_0 := V \setminus \mathbf{V}(X_0) \subseteq V$ is called an *affine open subset*, and $k[V]_0$ is the subalgebra of rational functions on V which are defined at all points of V_0. It is called the *affine coordinate ring* of V_0. If $\mathbf{I}(V)$ is generated by homogeneous polynomials f_i of degree d_i, then it is easy to see that $k[V]_0$ is the quotient of the polynomial ring $k[X_1/X_0,\dots,X_n/X_0]$ by the ideal $\mathbf{I}(V)_0$ of "dehomogenized" polynomials f_i/X_0^i.

We define $\dim(V) := \operatorname{trdeg}(k(V)/k)$. By (4.1.8), $\dim(V) \leq n$. What happens when $V = a$ is a point? By a linear change of variables, we may assume that $a = (1:0:\cdots:0)$. Then $\mathbf{I}(a) = (X_1,\dots,X_n)$, $k[V] \simeq k[X_0]$, and $k(V) = k$. So points have dimension zero. In fact, we have proved

Lemma 4.1.9. *If $a \in \mathbb{P}^n$ and X is any linear form with $X(a) \neq 0$, then the residue map $k[X_0,\dots,X_n] \to k[a]$ restricts to an isomorphism $k[X] \to k[a]$. In particular, points have dimension zero.* $\square$

We call V a *projective curve* if $\dim(V) = 1$, that is, if $k(V)$ is a function field as previously defined.

Lemma 4.1.10. *If $V \subseteq \mathbb{P}^2$ is a projective curve, then $\mathbf{I}(V) = (f)$ for some irreducible homogeneous polynomial f.*

Proof. Choose $f \in \mathbf{I}(V)$ homogeneous of minimal degree d. Since it is easy to see that the product of inhomogeneous polynomials is inhomogeneous, the minimality of d makes f irreducible. Put $x := X_1/X_0$ and $y := X_2/X_0$. Then $k(V) = k(x,y)$, and x and y satisfy the inhomogeneous polynomial $f_0 := f(1,X,Y)$, which is also irreducible. By (2.4.8) we have $k[V]_0 = k[x,y] = k[X,Y]/(f_0)$, which implies that $\mathbf{I}(V) = (f)$. □

4.2 Maps to $\mathbb{P}^n$

Given an arbitrary function field K/k, we ask if it can be realized as $k(V)$ for some projective curve $V \subseteq \mathbb{P}^n$. This turns out to be easy. In fact, we will construct a map from the points of K to $\mathbb{P}^n$ whose image is a projective curve V with $k(V) = K$. A more delicate question that then arises is whether this map is, with an appropriate definition, an embedding.

To answer the first question, let $\phi := (\phi_0, \phi_1, \dots, \phi_n) \subseteq K$ with $\phi_0 \neq 0$, and put $R := k[\phi_0, \dots, \phi_n]$. To avoid trivialities, assume that $k(\phi_1/\phi_0, \dots, \phi_n/\phi_0) \neq k$. Let $K' \subseteq K$ be the field of fractions of R. Define $\Phi : k[X_0, X_1, \dots, X_n] \to R[T]$ for some indeterminate T via

$$\Phi(X_i) := \phi_i T \quad (0 \leq i \leq n).$$

Give $R[T]$ the natural grading (in which all elements of R are homogeneous of degree zero). Then Φ is a map of graded k-algebras, so $I := \ker \Phi$ is a graded ideal. Indeed, I is just the set of all polynomial relations satisfied by the ϕ_i.

Since R is an integral domain, $R[T]$ is also an integral domain, and therefore so is $S := \operatorname{im} \Phi$. It follows that I is prime, and $V = \mathbf{V}(I)$ is a projective variety with $k[V]$ isomorphic to the graded k-subalgebra $S \subseteq R[T]$ generated by $\{\phi_i T \mid 0 \leq i \leq n\}$. By (4.1.8) Φ induces an isomorphism $k(V) \simeq k(\phi_1/\phi_0, \dots, \phi_n/\phi_0) \subseteq K'$. Since we chose the ϕ_i so that $k(V) \neq k$, we have $\operatorname{trdeg}(k(V)/k) = 1$, and therefore V is a projective curve. We will often abuse notation by identifying $k(V)$ with a subfield of K. In particular, if $k(\phi_1, \dots, \phi_n) = K$ and $\phi_0 = 1$, we get $k(V) = K$.

However, more is true. Let P be a point of K, let t_P be a local parameter at P, and put $e_P := -\min_i\{v_P(\phi_i)\}$. Then $t_P^{e_P}\phi_i \in \mathcal{O}_P$ for all i, and $t_P^{e_P}\phi_i \notin P$ for at least one i. It follows that $a_i := t_P^{e_P}\phi_i(P) \in k$ for all i, and $a_i \neq 0$ for at least one i. Abusing notation slightly, we define for every $P \in \mathbb{P}_K$,

$$\phi(P) := (a_0 : a_1 : \cdots : a_n) \in \mathbb{P}^n,$$

and we easily verify that $\phi(P)$ is independent of the choice of local parameter t_P. Moreover, we claim that if f and g are homogeneous of the same degree and $g(\phi(P))) \neq 0$, then

$$\frac{f(\phi(P))}{g(\phi(P))} = \frac{\Phi(f)}{\Phi(g)}(P). \tag{4.2.1}$$

Indeed, if f is homogeneous of degree d, then from the definition of Φ we have

$$\Phi(f) = f(\phi_0 X_0, \phi_1 X_0, \dots, \phi_n X_0) = X_0^d f(\phi_0, \dots, \phi_n),$$

whence

$$\frac{\Phi(f)}{\Phi(g)} = \frac{f(\phi_0,\ldots,\phi_n)}{g(\phi_0,\ldots,\phi_n)} = \frac{f(t^{e_P}\phi_0,\ldots,t^{e_P}\phi_n)}{g(t^{e_P}\phi_0,\ldots,t^{e_P}\phi_n)},$$

and evaluating both sides at P yields (4.2.1). This establishes a fundamental connection between rational functions defined on projective space and elements of the function field K defined on points of $\mathbb{P}_K$.

Taking $f \in \mathbf{I}(V)$ we conclude from (4.1.6) that $\operatorname{im}\phi \subseteq V$. To obtain the reverse inclusion, let $a \in V$ and define

$$\mathscr{O}_a := \left\{ \frac{\Phi(f)}{\Phi(g)} \in K \mid g(a) \neq 0 \right\},$$

$$P_a := \left\{ \frac{\Phi(f)}{\Phi(g)} \in \mathscr{O}_a \mid f(a) = 0 \right\}.$$

Because $a \in V$, the map $\mathscr{O}_a \to k$ given by $\Phi(f)/\Phi(g) \mapsto f(a)/g(a)$ is well-defined and has kernel P_a. Since the elements of $\mathscr{O}_a \setminus P_a$ are evidently units, $\mathscr{O}_a$ is a local subring of K with maximal ideal P_a, which is called the *local ring at a*. By (1.1.6) there is a prime P of K with $\mathscr{O}_a \subseteq \mathscr{O}_P$ and $P \cap \mathscr{O}_a = P_a$. If P is any prime of K with $\mathscr{O}_a \subseteq \mathscr{O}_P$ and $P \cap \mathscr{O}_a = P_a$, (4.2.1) shows that $\mathbf{I}(\phi(P)) \supseteq \mathbf{I}(a)$, whence $\phi(P) = a$ by (4.1.6). We have proved

Theorem 4.2.2. *Let K/k be a function field and let $(\phi_0,\ldots,\phi_n) \in K$ with $\phi_0 \neq 0$ and ϕ_i/ϕ_0 nonconstant for some i. For any point P of K, let t_P be a local parameter at P and put $e_P := -\min_i\{v_P(\phi_i)\}$. Then*

$$\phi(P) := (t_P^{e_P}\phi_0(P) : t_P^{e_P}\phi_1(P) : \cdots : t_P^{e_P}\phi_n(P)) \in \mathbb{P}^n$$

is well-defined, is independent of the choice of t_P, and

$$V := \{\phi(P) \mid P \in \mathbb{P}_K\}$$

is a projective curve with $k(V) = k(\phi_1/\phi_0,\ldots,\phi_n/\phi_0)$. Moreover, if $a \in V$ and $\mathscr{O}_a$ is the local ring at a, then

$$\phi^{-1}(a) = \{P \in P_K \mid \mathscr{O}_a \subseteq \mathscr{O}_P \text{ and } \mathscr{O}_P \cap \mathscr{O}_a = P_a\}. \quad \square$$

Thus, every ordered $(n+1)$-tuple of functions $\phi = (\phi_0,\ldots,\phi_n) \subseteq K$, with at least one ratio ϕ_i/ϕ_0 nonconstant, determines a map $\phi : \mathbb{P}_K \to V$ from the points of K onto the points of a projective curve $V \subseteq \mathbb{P}^n$. We call ϕ a *projective map*. Replacing the $\{\phi_i\}$ by $\{\phi_i' := y\phi_i\}$ for any fixed function $y \in K$ simply replaces e_P by $e_P{}' = e_P - v_P(y)$ and does not change the definition of $\phi(P)$ at all. We therefore put $\phi \sim \phi'$ if there is a function $y \in K$ with $\phi_i' = y\phi_i$ for all i. Note that in this case $\Phi(f)/\Phi(g)$ is unchanged for f and g homogeneous of degree d, because both numerator and denominator are multiplied by y^d. In particular, the local ring $\mathscr{O}_a$ depends only on the equivalence class of ϕ.

We say that ϕ is *effective* if the subspace $\langle\phi\rangle := \langle\phi_0,\ldots,\phi_n\rangle \subseteq K$ contains the constants. Note that $\phi_0^{-1}\phi$ is effective so that every projective map is equivalent to an effective one. For this reason, there is usually no reason to consider maps that are not effective, so unless explicitly indicated, we will assume that projective maps are effective. Indeed, we will frequently assume that $\phi_0 = 1$.

The simplest example of a projective map is a map to $\mathbb{P}^1$ of the form $\phi = (1,x)$ where x is any nonconstant function. Let $V := \text{im}\phi$. It is easy to see that the zeros of x map to $(1:0)$ and the poles of x map to $(0:1)$. Since the zeros of any homogeneous polynomial in two variables form a finite subset of $\mathbb{P}^1$, V is either all of $\mathbb{P}^1$ or a single point, so it must be all of $\mathbb{P}^1$. In particular, we have proved that as P ranges over all nonpoles of x, $x(P)$ ranges over all of k.

Since almost all of the integers e_P of (4.2.2) are zero, we can define the divisor

$$[\phi] := \sum_{P \in \mathbb{P}_K} e_P P,$$

and write

$$v_P(\phi) := e_P = -\min_i v_P(\phi_i),$$

where $\phi = (\phi_0, \ldots, \phi_n)$. If $\phi' = y\phi$, then $[\phi'] = [\phi] - [y]$, so equivalence classes of projective maps define divisor classes.

If ϕ_i is replaced by $\phi_i' := \alpha\phi_i$ or $\phi_i' := \phi_i + \alpha\phi_j$ for some scalar α and some $j \neq i$, it is easy to verify that $v_P(\phi') \geq v_P(\phi)$. Since the change of variable is invertible, it follows that $v_P(\phi') = v_P(\phi)$ and therefore the divisor $[\phi]$ depends only on the subspace $\langle\phi\rangle$ and not on the particular basis chosen. In particular, if ϕ is effective, we can choose a basis with $\phi_0 = 1$. Hence, effective maps have effective divisors. In general, if $(\phi_0', \ldots, \phi_n')$ is a different basis for $\langle\phi\rangle$, the resulting map ϕ' is obtained from ϕ by a linear change of variables in projective space.

Finally, if $(\phi_0, \phi_1, \ldots, \phi_n)$ is linearly dependent over k, then V lies in some hyperplane and we really have a map to $\mathbb{P}^{n-1}$. So we will always assume that the ϕ_i are linearly independent over k.

A projective map $\phi : \mathbb{P}_K \to V$ determines an embedding $\phi^* : k(V) \to K$ given by $\phi^*(f/g) = \Phi(f)/\Phi(g)$ in the previous notation, where f and g are homogeneous polynomials of the same degree. When ϕ^* is an isomorphism we call ϕ a *birational* map. If $\phi_1 : \mathbb{P}_{K_1} \to V$ and $\phi_2 : \mathbb{P}_{K_2} \to V$ are two birational maps to the same projective curve V, the fields K_1 and K_2 may be identified via the specific isomorphism $(\phi_1^*)^{-1}\phi_2^*$. If we begin with a projective curve $V \subseteq \mathbb{P}^n \setminus V(X_0)$, the *natural map* $\mathbb{P}_{k(V)} \to V$ is the projective map $\phi := (1, X_1/X_0, \ldots, X_n/X_0)$, which is clearly birational.

We should point out that the term "natural," although convenient, is somewhat fictitious here, in that it assumes a fixed choice of variables in $\mathbb{P}^n$. A "more natural," but perhaps less convenient, approach would be to allow a nonsingular linear change of variable in $\mathbb{P}^n$.

An interesting class of birational examples is obtained by letting $x \in K$ be a separating variable. Then $K/k(x)$ is generated by a single element y by (A.0.17), which means that the map $\phi := (1,x,y)$ is a birational map to $\mathbb{P}^2$. Such a map is called a *plane model*. We will study maps to $\mathbb{P}^2$ in detail in Section 4.5. However, the following is worth noting here.

Lemma 4.2.3. *Let $\phi : \mathbb{P}_K \to V \subseteq \mathbb{P}^n$ be an effective birational map. Then there exists a separating variable $x \in \langle\phi\rangle$, and for each such x there exists $y \in \langle\phi\rangle$ such that $(1,x,y)$ is birational.*

Proof. Since $K = k(\langle\phi\rangle)$, we have $\langle\phi\rangle \not\subseteq K^p$; hence there is an element $x \in \langle\phi\rangle$ such that $K/k(x)$ is separable by (2.4.6). In particular, there are only finitely many intermediate fields, all of which contain $\langle 1,x\rangle$ and none of which contain $\langle\phi\rangle$. By (A.0.14) applied to $\langle\phi\rangle/\langle 1,x\rangle$, there is an element $y \in \langle\phi\rangle$ that is disjoint from every intermediate field. □

When ϕ is not birational, we obtain the following.

Lemma 4.2.4. *Let $\phi : \mathbb{P}_K \to V$ be a projective map. If $K \supseteq K' \supseteq \phi^*(k(V))$ for some subfield K' of K, there is a uniquely determined projective map $\phi' : \mathbb{P}_{K'} \to V$ such that the diagram*

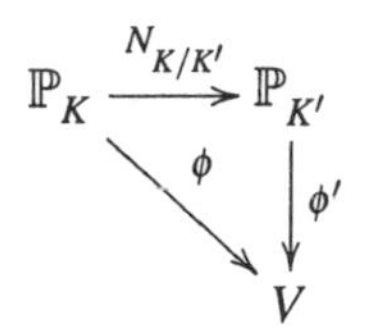

commutes. Moreover, we have

$$[\phi] = N^*_{K/K'}([\phi']). \tag{4.2.5}$$

Proof. If $Q \in \mathbb{P}_K$ and $P := Q \cap K'$, then $f(Q) = f(P)$ for any function $f \in K'$ by definition. Since all functions ϕ_i/ϕ_0 lie in K', we conclude that ϕ is constant on all divisors Q of P. Hence $\phi(Q) = \phi'(N_{K/K'}(Q))$, where ϕ' is the same ordered set of functions, now taken as elements of K'. Moreover, we have

$$\begin{aligned}\nu_Q(\phi) &= \min_i\{\nu_Q(\phi_i) \mid 0 \le i \le n\} = \min_i\{e(Q|P)\nu_P(\phi_i) \mid 0 \le i \le n\} \\ &= e(Q|P)\nu_P(\phi'),\end{aligned}$$

and (4.2.5) follows. □

When ϕ is effective, the divisor $[\phi]$ has the following important geometric interpretation. Namely, after a change of basis for $\langle\phi\rangle$, we may assume that $\phi_0 = 1$. Then for any $P \in \mathbb{P}_K$ the first co-ordinate of $\phi(P)$ is $t_P^{e_P}(P)$, which is zero if and only if $e_P > 0$. Thus, the divisor $[\phi]$ in this case describes the vanishing of the linear form X_0, or in other words, the intersection of the hyperplane $X_0 = 0$ with the image of ϕ. In particular, we see that there are only finitely many points of intersection.

For a general effective birational map $\phi = (\phi_0, \ldots, \phi_n)$, we have $1 = \sum_i a_i\phi$ for some $a_i \in k$. Then the divisor $[\phi]$ gives the vanishing of the linear form $\ell := \sum_i a_i X_i$. The zero set of this distinguished linear form is called the "hyperplane at infinity." Since we have

$$\phi^*\left(\frac{X_i}{\ell}\right) = \frac{\Phi(X_i)}{\Phi(\ell)} = \frac{\phi_i T}{T} = \phi_i,$$

we define $X_i^* = \phi_i$, and for any homogeneous polynomial $g(X_0,\ldots,X_n)$ of degree d, we define

$$g^* := \phi^*\left(\frac{g}{\ell^d}\right) = g(\phi_0,\ldots,\phi_n) \in K.$$

Note that g^* depends on the choice of hyperplane at infinity, and if we replace ℓ by ℓ_1, g^* is replaced by g^*/ℓ_1^*.

A point $\phi(P)$ lies on the hyperplane at infinity precisely when $e_P > 0$, that is, when some ϕ_i has a pole at P. We can always replace ϕ by an equivalent effective map ϕ' with $v_P([\phi']) = 0$. When this happens, we say that ϕ is *normalized at* P. We will sometimes abuse notation by saying that a point P is "infinite" when $\phi(P)$ lies on the hyperplane at infinity, and "finite" when it does not.

Replacing ϕ by an equivalent effective map amounts to choosing a different hyperplane to lie at infinity. To put the hyperplane $\sum_i a_i X_i = 0$ at infinity, let $y := \sum_i a_i \phi_i \in \langle\phi\rangle$, and replace ϕ by $\phi' := y^{-1}\phi$. In particular, it follows that every hyperplane meets V in a finite set. More generally, suppose $g(X_0,\ldots,X_n)$ is homogeneous of degree d with $g \notin \mathbf{I}(V)$. Then $g^* \neq 0$ and if $\phi(P) \in \mathbf{V}(g)$ then either $X_0(\phi(P)) = 0$ or $g^*(P) = 0$. Since g^* has only finitely many zeros, we conclude that $|V \cap \mathbf{V}(g)|$ is finite. This implies that every closed set that does not contain V meets V in a finite set. Finally, given $a \in V$ we can choose a hyperplane at infinity containing a. Then $\phi(P) = a$ implies $v_P(\phi) > 0$. We have proved

Lemma 4.2.6. *Let $\phi : \mathbb{P}_K \to V$ be a projective map. Then every closed set that does not contain V meets V in a finite set. Moreover, $\phi^{-1}(a)$ is finite for every $a \in V$.* □

For $a \in V$ and $\phi : \mathbb{P}_{k(V)} \to V$ the natural map, we say that the hyperplane $\mathbf{V}(X_0)$ meets V at a with multiplicity

$$\sum_{P \in \phi^{-1}(a)} v_P(\phi).$$

More generally, if $g(X_0,\ldots,X_n)$ is a homogeneous polynomial of degree d in $n+1$ variables, we define the *intersection divisor* of V and the closed set $\mathbf{V}(g)$ as

$$[\phi]_g := d[\phi] + [g^*].$$

Note that g need not be irreducible here. Indeed, if $g =: \prod_i g_i^{e_i}$ where the g_i are irreducible, then $\mathbf{V}(g)$ is a union of projective varieties (called *hypersurfaces*) $\mathbf{V}(g_i)$ counted with multiplicity e_i, and $[g^*] = \sum_i e_i [g_i^*]$. Thus,

$$[\phi]_g = \sum_i e_i [\phi]_{g_i}.$$

It appears that $[\phi]_g$ depends on the choice of a hyperplane at infinity, but in fact it does not. If ℓ is any linear form, we can put the hyperplane $\mathbf{V}(\ell)$ at infinity by taking $\phi' := \ell^{*-1}\phi$. Then the new dehomogenization of g is $g' := \ell^{*-d} g^*$ and it is straightforward to check that $[\phi']_g = [\phi]_g$.

Let $\phi = (X_i/X_0 \mid 0 \leq i \leq n)$ be the natural map from the points of $k(V)$ to V. We define the *intersection multiplicity* of the closed set $\mathbf{V}(g)$ with the curve V at the point $a \in V \cap \mathbf{V}(g)$ to be

$$\mu_a(\mathbf{V}(g)) := \sum_{P \in \phi^{-1}(a)} v_P([\phi]_g).$$

It is not entirely obvious that these multiplicities are nonnegative, a property which is implied by the terminology. We will prove this below.

We define the *degree* of a projective map ϕ to be $\deg \phi := \deg[\phi]$, and the degree of a projective curve V to be the degree of the natural map. For a plane curve we will show in (4.5.4) that the degree is what it should be, which is the degree of the defining irreducible homogeneous polynomial generating $\mathbf{I}(V)$ (see (4.1.10)).

Lemma 4.2.7. *Given any finite set of points in $\mathbb{P}^n$, there exists a hyperplane not containing any of them.*

Proof. The vanishing of the linear form $\sum_i a_i X_i$ at a point forces $(a_0, \ldots, a_n)$ to lie in an n-dimensional subspace of k^{n+1}. So the lemma follows from (A.0.14). □

Theorem 4.2.8 (Bézout). *Let $V \subseteq \mathbb{P}^n$ be a projective curve of degree d with $K := k(V)$ and natural map ϕ. Suppose that $g(X_0, \ldots, X_n) \notin \mathbf{I}(V)$ is a (possibly reducible) homogeneous polynomial of degree e. Then there exists a hyperplane containing no points of $V \cap \mathbf{V}(g)$, and for any such hyperplane $\mathbf{V}(\ell)$ we have $[\phi]_g = [\phi^*(g/\ell^e)]_0$. In particular, $[\phi]_g \geq 0$, and with the definition of intersection multiplicity given above, the closed set $\mathbf{V}(g)$ meets V at de points.*

Proof. Since $g \notin \mathbf{I}(V)$ we have $g^* \neq 0$. By (4.2.6) $|V \cap \mathbf{V}(g)|$ is finite, so the lemma yields a hyperplane $\mathbf{V}(\ell)$ that does not meet $V \cap \mathbf{V}(g)$. Because $[\phi]_g$ is independent of the choice of hyperplane at infinity, we can replace ϕ by the equivalent effective map $\ell^{*-1}\phi$ and change notation so that there are no points of $V \cap \mathbf{V}(g)$ at infinity. Thus, if $P \in \mathbb{P}_K$ and $g(\phi(P)) = 0$, then $g^*(P) = 0$. However, we need to get the multiplicities right. Thus, we want to prove that $[\phi]_g = [g^*]_0$, or what is the same thing, that $e[\phi] = [g^*]_\infty$.

Since $g^* = g(1, \phi_1, \ldots, \phi_n)$ and g is a polynomial, it follows that if $P \in \mathbb{P}_K$ is a pole of g^*, then $X_0(\phi(P)) = 0$. Let P be any point of $\mathbb{P}_K$ with $X_0(\phi(P)) = 0$. Choosing notation so that $X_1(\phi(P)) \neq 0$, we have

$$g^* = \phi_1^e \phi^* \left(\frac{g(X_0, \ldots, X_n)}{X_1^e} \right).$$

By our choice of hyperplane at infinity, $g(\phi(P)) \neq 0$. We conclude that $v_P(g^*) = e v_P(\phi_1)$. On the other hand, $X_1(\phi(P)) \neq 0$ implies that

$$\min_i \{v_P(\phi_i)\} = v_P(\phi_1).$$

Therefore, we have $v_P(g^*) = -e v_P(\phi)$ for every point $P \in \mathbb{P}_K$ with $X_0(\phi(P)) = 0$. This implies that $[g^*]_\infty = e[\phi]$, and thus $[\phi]_g = [g^*]_0$, as required. Since g^* vanishes at a point $P \in \mathbb{P}_K$ if and only if $\phi(P) \in V \cap \mathbf{V}(g)$, the number of points of intersection counting multiplicities as defined above is

$$\deg[g^*]_0 = \deg[g^*]_\infty = de. \quad \square$$

In particular, we see that every hyperplane meets V in $\deg V$ points, justifying the definition of $\deg V$. In general, the intersection theory of projective varieties is a delicate and complicated subject. The interested reader may want to see [8].

Before proceeding, we digress briefly to discuss the connection of the above ideas with other closely related notions. When ϕ is effective, the set of all divisors $\{[\phi]_f \mid f \in \langle\phi\rangle\}$ is known in the literature as a "base-point-free linear system." These divisors of course determine the subspace $\langle\phi\rangle$ and thus the map ϕ up to a choice of coordinates in projective space. It is immediate from our definitions that $\langle\phi\rangle \subseteq L([\phi])$. If $\langle\phi\rangle = L([\phi])$, the corresponding linear system is called a "complete linear system." We will discuss the corresponding projective maps below.

Since every curve $V \subseteq \mathbb{P}^n$ arises as the image of a natural map ϕ from the points of its function field to $\mathbb{P}^n$, we can think of a projective curve as being naturally "parametrized" by the points of its function field, via the map ϕ. We know that ϕ is surjective, but is it bijective? This would certainly follow if it happened that the local ring $\mathcal{O}_a$ were a valuation ring for every $a \in V$, because then we would have $\mathcal{O}_a = \mathcal{O}_P$ for some $P \in \mathbb{P}_K$ uniquely determined by a. This brings us to the subject of the next section.

4.3 Projective Embeddings

Let K be a function field and let $\phi : \mathbb{P}_K \to V$ be a projective map with $\phi(P) = a$ for some point $P \in \mathbb{P}_K$. We say that ϕ is *nonsingular* at P (or sometimes, at a) if $\mathcal{O}_a = \mathcal{O}_P$. Some authors use the term *smooth* here. Note that in particular, ϕ will be one-to-one at such a point P. Moreover, since $\mathcal{O}_a \subseteq k(V)$ and the field of fractions of $\mathcal{O}_P$ is K, we have

Lemma 4.3.1. *If the projective map $\phi : \mathbb{P}_K \to k(V)$ is nonsingular at any point, then ϕ is birational.* $\square$

We call ϕ an *embedding* if it is nonsingular at every point. Suppose that $V \subseteq \mathbb{P}^n$ is a projective curve with function field K. We say that V is *nonsingular* at a point $a \in V$ if the natural map $\phi : \mathbb{P}_K \to V$ is nonsingular at a, and that V is nonsingular if it is nonsingular at every point. Evidently, V is nonsingular at a if and only if $\mathcal{O}_a$ is a discrete valuation ring.

Our major tool for the study of singularities is the following result.

Theorem 4.3.2. *Let $\phi : \mathbb{P}_K \to V \subseteq \mathbb{P}^n$ be a birational map and let $a \in V$. Let R_a be the integral closure of $\mathcal{O}_a$ in $k(V)$. Then every ideal I of $\mathcal{O}_a$ contains a nonzero ideal of R_a and R_a/I is a finite-dimensional vector space over k. In particular, every nonempty set of ideals of $\mathcal{O}_a$ has a maximal element.*

Proof. By (4.2.6) and (1.1.8), R_a is the intersection of finitely many valuation rings $\mathscr{O}_{P_1},\ldots,\mathscr{O}_{P_m}$ of $k(V)$, where $\{P_1,\ldots,P_m\} = \phi^{-1}(a)$. Let $\mathscr{V}$ denote the corresponding set of valuations. Referring to (1.1.17), we see that $R_a = K(\mathscr{V};0)$, and that every nonzero ideal of R_a has finite codimension in R_a. It remains to show that every nonzero ideal of $\mathscr{O}_a$ contains a nonzero ideal of R_a.

We claim that $\mathscr{O}_a \supseteq K(\mathscr{V};N)$ for some suitably large integer N, where by abuse of notation, $N(v) = N$ for all $v \in \mathscr{V}$. To see this, we first make a careful change of variable in $\mathbb{P}^n$. Since $\mathscr{O}_a$ depends only on the equivalence class of ϕ, we may assume that ϕ is effective and that a does not lie on the hyperplane at infinity. Since $k(\langle\phi\rangle) = K$, $K^p \cap \langle\phi\rangle$ is a proper subspace of $\langle\phi\rangle$ containing k, and therefore by (A.0.14) applied to $\langle\phi\rangle$, we can choose coordinate functions $1,x_1,\ldots,x_n$ such that x_i is a separating variable for $i \geq 1$. Using these coordinate functions, let $a =: (1 : a_1 : \cdots : a_n)$. Since k is algebraically closed and $x_i \notin K^p$ (see (2.4.6)), $x_i - a_i$ is also a separating variable, so replacing x_i by $x_i - a_i$ for $1 \leq i \leq n$ and changing notation again, we have $a = (1 : 0 : \cdots : 0)$.

For each i, let $\mathscr{V}_i$ be the set of all valuations of K that are positive at x_i. Then $\mathscr{V} = \cap_i \mathscr{V}_i$. Note that $K(\mathscr{V}_i;0)$ is the integral closure in K of $\mathscr{O}_{(x_i)}$, the localization of $k[x_i]$ at x_i. By (4.2.3) there is an element $y_i \in \langle\phi\rangle \subseteq \mathscr{O}_a$ with $K = k(x_i,y_i)$. We can write

$$\sum_{j=0}^{n} f_{ij}(x_i)y_i^j = 0,$$

where $f_{ij} \in k[x_i]$ and $f_{in} \neq 0$. Put $y_i' := f_{in}y_i$. Then y_i' is integral over $k[x_i]$ and $K = k(x_i,y_i)$ for all i.

Since we have $x_i = \phi^*(X_i/X_0) \in \mathscr{O}_a$ for all i, it follows that $\langle\phi\rangle \subseteq \mathscr{O}_a$, whence $k[x_i] \subseteq \mathscr{O}_a$ for all i. In particular, $y_i' \in \mathscr{O}_a$ for all i. Moreover, any polynomial in $k[x_i]$ with nonzero constant term is a unit in $\mathscr{O}_a$ and therefore $\mathscr{O}_{(x_i)} \subseteq \mathscr{O}_a$. We conclude that $\mathscr{O}_{(x_i)}[y_i'] \subseteq \mathscr{O}_a$, whence (3.3.10) shows that $\mathscr{O}_a$ contains an ideal of $K(\mathscr{V}_i;0)$ for all i. Thus, for a suitably large value of N we have $K(\mathscr{V}_i;N) \subseteq \mathscr{O}_a$ for all i by (1.1.17). Now (1.1.18) yields $K(\mathscr{V};N) \subseteq \mathscr{O}_a$, and therefore any nonzero ideal I of $\mathscr{O}_a$ contains the non-zero ideal $IK(\mathscr{V};N)$ of R_a. □

We can now obtain a key equivalence.

Corollary 4.3.3. *Let $\phi : \mathbb{P}_K \to V \subseteq \mathbb{P}^n$ be a projective map and let $a \in V$. In order that ϕ be nonsingular at a it is necessary and sufficient that $\mathscr{O}_a$ be contained in a unique valuation ring $\mathscr{O}_P$ of K and contain a local parameter t at P.*

Proof. The necessity of the conditions being obvious, we argue that they are sufficient. Since $\mathscr{O}_P$ is the unique valuation ring of K containing $\mathscr{O}_a$, it is the integral closure of $\mathscr{O}_a$ in K by (1.1.8). By (4.3.2) we have $P^m \subseteq \mathscr{O}_a$ for some m.

Let t be a local parameter at P with $t \in \mathscr{O}_a$. Since $k + P = \mathscr{O}_P$, multiplication by t^i yields

$$kt^i + P^{i+1} = P^i$$

for all $i \geq 0$. It follows that $\mathcal{O}_P/P^m$ is generated as a k-algebra by $t + P^m$, and we get $\mathcal{O}_a/P^m = \mathcal{O}_P/P^m$. Since $\mathcal{O}_a \supseteq P^m$, we have $\mathcal{O}_a = \mathcal{O}_P$. □

The two conditions of (4.3.3) correspond to two different types of singularities. If $\mathcal{O}_a$ is contained in more than one valuation ring, the map fails to be one-to-one at a. If $\mathcal{O}_a$ does not contain a local parameter, the tangent line is not well-defined at a. For more on the tangent line, see Section 4.5.

We now proceed to our second major equivalence. In order to state this, we need

Theorem 4.3.4 (Hilbert Basis Theorem). *If every ideal of the commutative ring R is finitely generated, then the same is true for $R[X]$. In particular, every ideal of $k[X_0, \ldots, X_n]$ is finitely generated.*

Proof. Let I be an ideal in $R[X]$. To show that I is finitely generated, we choose a sequence of polynomials $f_i \in I$ of degree d_i and leading coefficient a_i $(i = 0, 1, \ldots)$ as follows. Choose $f_0 \neq 0$ of minimal degree. If $I_{i-1} := (f_0, f_1, \ldots, f_{i-1}) \subsetneq I$, choose $f_i \in I \setminus I_{i-1}$ of minimal degree; otherwise, take $f_i = f_{i-1}$. Note that $d_i \leq d_{i+1}$ for all i. For some i, there will exist elements $r_0, \ldots, r_{i-1} \in R$ such that

$$a_i = \sum_{j=0}^{i-1} r_j a_j,$$

because the ideal of R generated by the a_i $(i = 0, 1, \ldots)$ is finitely generated. Then the polynomial

$$f = f_i - \sum_{j=0}^{i-1} r_j X^{d_i - d_j} f_j$$

has lower degree than f_i, and if $f_i \in I \setminus I_{i-1}$, so is f. By our choice of f_i, we conclude that $I = I_i$, as required. □

In the classical case over the complex numbers, the implicit function theorem gives a criterion for the zero set of a finite number of polynomials to be smooth at a point, namely, that the matrix of partial derivatives have maximal rank at that point. This notion generalizes to any field k as follows. Let $V \subset \mathbb{P}^n$ be a projective variety. Then $\mathbf{I}(V)$ is finitely generated by (4.3.4), and by using (4.1.1) we can in fact find generators $\{f_1, \ldots, f_r\}$ where f_i is homogeneous of degree d_i. Choosing notation as usual so that $X_0(a) \neq 0$, the rational functions

$$f_{ij} := \frac{\partial f_i / \partial X_j}{X_0^{d_i - 1}} \quad (1 \leq i \leq r,\ 1 \leq j \leq n)$$

are defined at a, and we have

Theorem 4.3.5. *Let $V \subseteq \mathbb{P}^n$ be a projective curve with function field K and let $a \in V$ with notation chosen so that $X_0(a) \neq 0$. In order that V be nonsingular at a it is necessary and sufficient that the matrix $f_{ij}(a)$ have rank $n - 1$.*

Proof. In order to make calculations, we are going to work in the dehomogenized affine ring. If $f \in k[X_0,\dots,X_n]$ is homogeneous of degree d, we put $\hat{f} := f/X_0^d \in A := k[\hat{X}_1,\dots,\hat{X}_n]$, where $\hat{X}_i := X_i/X_0$. Then

$$f_{ij} = \frac{\partial \hat{f}_i}{\partial \hat{X}_j}.$$

Let $a =: (1 : a_1 : \cdots : a_n)$. Since the zero set of the ideal generated by $\{e_j := X_j - a_j X_0 \mid 1 \le j \le n\}$ is just a, it follows from (4.1.4) that $\mathbf{I}(a)$ is generated by the e_j. Then the dehomogenization of $\mathbf{I}(a)$ is $I_a := (\hat{e}_1,\dots,\hat{e}_n)$, a maximal ideal of A. Since $A/I_a = k$, the A-module I_a/I_a^2 is spanned as a k-vector space by the $\hat{e}_j$. For any polynomial $g \in A$ define

$$g_j := \frac{\partial g}{\partial \hat{X}_j} \quad (1 \le j \le n),$$

and put $\theta(g) := (g_1(a),\dots,g_n(a)) \in k^n$. Since $\theta(\hat{e}_j)$ is the jth standard basis vector for k^n, we have $\theta(I_a) = k^n$. From the product rule for partial derivatives we check that $\theta(I_a^2) = 0$, which implies that $\dim_k I_a/I_a^2 = n$ and thus that $I_a^2 = \ker\theta$.

Let $I_V = (\hat{f}_1,\dots,\hat{f}_r) \subseteq A$ be the ideal of A corresponding to V. Using the product rule again, we see that the subspace $\theta(I_V)$ is spanned by the columns of the matrix $f_{ij}(a)$ because $\hat{f}_i(a) = 0$ for all i. It follows that the condition of the theorem is equivalent to the condition

$$\operatorname{codim}_k \theta(I_V) = 1.$$

We have $I_a \supseteq I_V + I_a^2 \supseteq I_a^2$, so a further equivalent condition is

$$\dim_k I_a/(I_V + I_a^2) = 1.$$

We are now in a position to restrict functions to V, or in other words, to reduce modulo I_V. The algebra A/I_V is just the affine coordinate ring $k[V]_0 \subseteq k(V)$ defined in the previous section. Put $M_a := I_a/I_V \subseteq k[V]$. Then we are trying to prove that

$$V \text{ is nonsingular at } a \text{ if and only if } \dim M_a/M_a^2 = 1.$$

Tracing through the definitions, we see that $\mathcal{O}_a$ is just the localization of $k[V]_0$ at M_a, so it is easy to see that the vector spaces M_a/M_a^2 and P_a/P_a^2 are isomorphic. We are thus reduced to proving

$$V \text{ is nonsingular at } a \text{ if and only if } \dim P_a/P_a^2 = 1.$$

Since one implication is obvious, we will suppose that $\dim P_a/P_a^2 = 1$ and argue that $\mathcal{O}_a$ is a valuation ring. Choose an element $t \in P_a \setminus P_a^2$. Then $P_a = \mathcal{O}_a t + P_a^2$, and therefore the $\mathcal{O}_a$-module $N := P_a/\mathcal{O}_a t$ satisfies $P_a N = N$. Since $\dim_k N < \infty$ by (4.3.2), we have $P_a = \mathcal{O}_a t$ by Nakayama's Lemma (1.1.5).

For any $x \in P_a$ we can thus write $x = tx_1$ for some $x_1 \in \mathcal{O}_a$. If $x_1 \in P_a$ write $x_1 = tx_2$. Continuing in this way we eventually obtain either $x_n \notin P_a$ for some n, or we get an infinite properly increasing chain of principal ideals $\mathcal{O}_a x \subseteq \mathcal{O}_a x_1 \subseteq \cdots$, contrary to (4.3.2). It follows that $x = t^n u$ for some unit $u \in \mathcal{O}_a$, and therefore $\mathcal{O}_a$ is a valuation ring by (1.1.10). Since the field of fractions of $\mathcal{O}_a$ is $k(V)$, the natural map $\mathbb{P}_{k(V)} \to V$ is nonsingular at a, as required. □

We remark that the only application of (4.3.5) in this book is to the case $n = 2$. Here, $I(V)$ is generated by a single irreducible homogeneous polynomial $f(X_0,X_1,X_2)$ by (4.1.10). If we put $x := X_1/X_0$ and $y := X_2/X_0$, the affine equation is $\hat{f}(x,y) = f(1,x,y) = 0$. The theorem says that V is singular at a point $(1 : a : b)$ if and only if $\hat{f}_x(a,b) = \hat{f}_y(a,b) = 0$. If we denote the partial derivatives of $f(X_0,X_1,X_2)$ by f_0, f_1, f_2 respectively, this is equivalent to saying that $f_1(1,a,b) = f_2(1,a,b) = 0$. It is easy to check that the condition $f_0(1,a,b) = 0$ is redundant. For points at infinity, with (say) $X_1 \neq 0$, the conditions amount to $f_0(0,a,b) = f_2(0,a,b) = 0$. So to find all singularities, we have

Corollary 4.3.6. *Let $V \subseteq \mathbb{P}^2$ with defining polynomial $f(X_0,X_1,X_2)$. Let $f_i := \partial f/\partial X_i$ for $i = 0,1,2$. Then V is singular at $(a : b : c) \in V$ if and only if $f_i(a,b,c) = 0$ for each i.* □

Corollary 4.3.7. *A projective curve has only finitely many singularities. A birational map is nonsingular at almost all points.*

Proof. The two statements are evidently equivalent. Let $\phi = (1, \phi_1, \ldots, \phi_n) : \mathbb{P}_K \to V$ be a birational map. We consider first the special case $n = 2$. By (4.1.10) we have $V = V(f)$ for some irreducible polynomial $f(X_0,X_1,X_2)$, and by (4.3.6) the singularities are the closed set $V(f) \cap V(f_0, f_1, f_2)$. Since the f_i have degree less than $\deg(f)$, they are not divisible by f and therefore do not vanish on V. So the set of singularities is finite by (4.2.6).

In the general case, we note that by (4.2.3) there exists a birational map $(1,x,y)$ with $\{x,y\} \subseteq \langle\phi\rangle$. Extending $(1,x,y)$ to a basis for $\langle\phi\rangle$ and changing notation, we may assume that $\phi' := (1, \phi_1, \phi_2)$ is birational. Let $P \in \mathbb{P}_K$ and put $a := \phi(P)$ and $a' := \phi'(P)$. At this point, some care is needed. If $a = (1 : a_1 : a_2 : \cdots : a_n)$ is a finite point, then $a' = (1 : a_1 : a_2)$, and it is clear from the definitions that $\mathcal{O}_{a'} \subseteq \mathcal{O}_a \subseteq \mathcal{O}_P$. This implies that if ϕ' is nonsingular at P so is ϕ. If, however, some coordinate function has a pole at a, we might have $a = (0 : 0 : 0 : a_3 : \cdots : a_n)$, in which case a' has no obvious relation to a. Fortunately, there are only finitely many points at infinity, and therefore since ϕ' has only finitely many singularities, so does ϕ. □

As an example, we apply (4.3.5) to the function field $K = \mathbb{C}(x,y)$ over the complex numbers where $y^2 = x^3 - x$. The functions $(1,x,y)$ define a map to $\mathbb{P}^2$ whose image is the plane curve $V := \mathbf{V}(X_0X_2^2 - X_1^3 + X_0^2X_1)$. In this case, the matrix f_{ij} above is given in homogeneous form by

$$(X_0^2 + 2X_0X_1, -3X_1^2 + X_0^2, 2X_0X_1),$$

and by (4.3.5), V will be singular only at points a at which two components vanish. An easy calculation shows that none of these points lie on the curve.

However, since $x = y^2/(x^2-1)$, the functions $\{x^2, y\}$ also generate K. Using the functions $(1, x^2, y)$ instead of $(1, x, y)$, we obtain the plane curve $V_1 := \mathbf{V}(X_2^4 - X_0X_1^3 + 2X_0^2X_1^2 - X_0^3X_1)$. Calculating as above, however, we find a singularity at $(1:1:0)$. What has happened is that each point $(1:a_1:a_2)$ on V has been mapped to the point $(1:a_1^2:a_2)$ on V_1. In particular, the two distinct points $(1:1:0)$ and $(1:-1:0)$ have been identified, creating what is called a double point on V_1. In the terminology of (4.3.3), $\mathscr{O}_{(1:1:0)}$ is not contained in a unique valuation ring of K.

Another type of singularity is illustrated by the map $(1, x, yx)$, which yields the plane curve $V_2 := \mathbf{V}(X_0^3X_2^2 - X_1^5 + X_1^3X_0^2)$. Computing partial derivatives as above, we find that V_2 has a singularity at the origin, but rather than a double point, we get a so-called cusp. In the terminology of (4.3.3), $\mathscr{O}_{(1:0:0)}$ does not contain a local parameter[1].

To see this, put $r := X_1/X_0$ and $s := X_2/X_0$. Since $s^2 = r^5 - r^3$, we have $2v_P(s) = 3v_P(r)$, where P is the point of K mapping to $(1:0:0)$. Since $v_P(s)$ and $v_P(r)$ are both positive, each must be at least two. Every element of $\mathbb{P}_a$ has the form $u := f(1,r,s)/g(1,r,s)$ with f and g homogeneous of the same degree and $f(1,0,0) = 0 \neq g(1,0,0)$. This means that $v_P(u) = v_P(f(1,r,s)) \geq 2$.

The preceeding example leads directly to the following result.

Lemma 4.3.8. *Let ϕ be an effective projective map with $\phi(P) = a$ and assume that ϕ is normalized at P. Then $\mathscr{O}_a$ contains a local parameter at P if and only if $\langle\phi\rangle$ contains a local parameter at P.*

Proof. Make a linear change of basis so that $a = (1:0:\cdots:0)$. This amounts to choosing a basis $(1, \phi_1, \ldots, \phi_n)$ for $\langle\phi\rangle$ such that $\phi_i(P) = 0$ for $i \geq 1$. Now we can write

$$\sum_i a_i\phi_i = \phi^*\left(\frac{\sum_i a_iX_i}{X_0}\right),$$

which shows that $\langle\phi\rangle \subseteq \mathscr{O}_a$, and thus one implication is trivial. Conversely, we are assuming that there are homogeneous polynomials f and g of degree d with $g(a) \neq 0$ such that $v_P(f^*/g^*) = 1$. Since $g(a) \neq 0$, we obtain $v_P(g^*) = 0$ and $v_P(f^*) = 1$. Now, f^* is a polynomial in $\{\phi_1, \ldots, \phi_n\}$ that vanishes at P, so we can write $f^* = \sum_{i\geq 1} f_i^*$, where f_i^* is homogeneous of degree i in $\phi_1, \ldots, \phi_n$. Clearly, $f_1^* \in \langle\phi\rangle$. Since the ϕ_i vanish at P for $i \geq 1$, we have $v_P(f_i^*) \geq i$, and thus we must have $f_0^* = 0$ and $v_P(f_1^*) = 1$. □

Corollary 4.3.9. *Let $V \subseteq \mathbb{P}^n$ be a projective curve. Then V is nonsingular at a if and only if some hyperplane has intersection multiplicity 1 at a.*

Proof. Choose notation so that $X_0(a) \neq 0$ and let $\phi : \mathbb{P}_{k(V)} \to V$ be the natural map. Suppose V is nonsingular at a and let $\phi(P) = a$. Since ϕ is effective and a

---

[1] Geometrically, this means that the curve has no tangent line at the origin, as we will show in Section 4.5.

is not at infinity, there is a function $f = \sum_i a_i \phi_i$ with $v_P(f) = 1$ by (4.3.8), and therefore $\mu_a(\mathbf{V}(\sum_i a_i X_i)) = 1$.

Conversely, if there is a hyperplane with intersection multiplicity 1 at a, then choosing notation so that $X_0(a) \neq 0$, there is a linear form f with

$$\sum_{P \in \phi^{-1}(a)} v_P(f/X_0) = 1.$$

Since $v_P(f/X_0) > 0$ for every $P \in \phi^{-1}(a)$, (4.3.3) completes the proof. □

We next consider the problem of trying to actually construct nonsingular projective maps. In (4.2.3), we constructed birational maps to $\mathbb{P}^2$, so we might wonder whether we can find a nonsingular map to $\mathbb{P}^2$. The answer in general is no, as we will see later in (4.5.17). However, we can prove

Theorem 4.3.10. *For any $x \in K$, there exists $y \in K$ such that $\phi := (1,x,y)$ is birational and $\langle \phi \rangle$ contains a local parameter at every point $P \in \mathbb{P}_K$. Thus, the only singularities of ϕ are multiple points.*

Proof. This is an easy consequence of the weak approximation theorem. Let S be the (finite) set of all points of K that are either ramified over $k(x)$ or lie over (x^{-1}). Let $x - a$ be a point of $k(x)$ that is unramified in K and let $P_1, \ldots, P_n$ be the set of all points of K lying over it. Choose distinct elements $a_1, \ldots, a_n \in k$. By (1.1.16) there exists $y \in K$ such that $v_P(y) = 1$ for all $P \in S$, and $v_{P_i}(y - a_i) > 0$ for all i.

For $P \in S$, y is a local parameter at P. For $P \notin S$ we have $x(P) = b$ for some $b \in k$ because P does not lie over x^{-1}, and $v_P(x - b) = 1$ because P is unramified. Finally, since $\phi(P_i) = (1 : a : a_i)$, (4.2.4) implies that ϕ is birational, because it has distinct values at all the divisors of $(x - a)$. □

If we are willing to go to $\mathbb{P}^3$, the strong approximation theorem yields an embedding.

Theorem 4.3.11. *Let $\phi := (1,x,y)$ be a birational map to $\mathbb{P}^2$ with no singularities at infinity. Then there exists $z \in K$ such that $(1,x,y,z)$ is nonsingular.*

Proof. Let $\{P_1, \ldots, P_n\}$ be the set of singularities of ϕ. Choose distinct elements $a_1, \ldots, a_n \in k$, and let P_∞ be a nonsingular finite point of ϕ. By (2.2.13) there exists $z \in K$ such that $v_{P_i}(z - a_i) = 1$ for all i, and the only pole of z is at P_∞. Put $\hat{\phi} := (1,x,y,z)$. By (4.3.3), we need to show that $\hat{\phi}$ is one-to-one and contains a local parameter at every point.

For $P \notin \{P_1, \ldots, P_n\}$, $\langle 1,x,y \rangle$ contains a local parameter at P, and $z - a_i$ is a local parameter at P_i for all i. Thus, it remains to show that $\hat{\phi}$ is one-to-one. By our construction, $\hat{\phi}(P_\infty) = (0:0:0:1)$ is the unique point at which the first three coordinates vanish because P_∞ is the unique pole of z, and it is a finite point of ϕ. If P is any other point with $\hat{\phi}(P)$ at infinity, then $\hat{\phi}(P) = (0:a:b:0)$ for some $a, b \in k$. These images are distinct, because ϕ has no singularities at infinity. At finite points P of $\hat{\phi}$ at which ϕ is nonsingular, the first three coordinates of $\hat{\phi}(P)$

are distinct, while if $P = P_i$ for any i, the fourth coordinate is a_i by construction. It follows that $\hat{\phi}$ is nonsingular. □

We remark that every plane map is equivalent to one with no singularities at infinity because by (4.2.7) we can choose a hyperplane at infinity that avoids the finitely many singular points.

We next turn to a characterization of singularities in terms of subspaces of $\langle\phi\rangle$. It is immediate from our definitions that $\langle\phi\rangle \subseteq L([\phi])$. Given a nonnegative divisor D, define $L_\phi(D) := \langle\phi\rangle \cap L([\phi]-D)$. Then $\langle\phi\rangle = L_\phi(0)$ and if $D_1 \leq D_2$ then $L_\phi(D_1) \supseteq L_\phi(D_2)$. In particular, at each point $P \in \mathbb{P}_K$ we have the important filtration

$$\langle\phi\rangle = L_\phi(0) \supseteq L_\phi(P) \supseteq L_\phi(2P) \supseteq \cdots, \tag{4.3.12}$$

which we call the *osculating filtration at P*. We will study this filtration in detail in Section 4.4. For now, we have $\dim L_\phi(nP)/L_\phi((n+1)P) \leq 1$ from (2.1.10). Moreover, we have equality if and only if there is a function $f \in \langle\phi\rangle$ with $\nu_P(f) = n - \nu_P(\phi)$. By definition there is always a function $f \in \langle\phi\rangle$ with $\nu_P(f) = \nu_P(\phi)$, so $L_\phi(P)$ is always a hyperplane of $\langle\phi\rangle$.

Note that the subspace $L_\phi(D)$ does not depend on any particular choice of basis, and that if $\phi' = y\phi$ is an equivalent map, then $yL_\phi(D) = L_{\phi'}(D)$. In fact, since $L_\phi(P)$ is a point of the dual projective space $\langle\phi\rangle^*$, the map $P \to L_\phi(P)$ is a projective map that is easily seen to be equivalent to ϕ. This observation can be used to develop a coordinate-free treatment of projective maps.

Corollary 4.3.13. *Suppose that K is a function field, $\phi : \mathbb{P}_K \to V \subseteq \mathbb{P}^n$ is a projective map, and $P \in \mathbb{P}_K$. Then ϕ is nonsingular at P if and only if for every point $Q \in \mathbb{P}_K$ we have* $\operatorname{codim} L_\phi(P+Q) = 2$.

Proof. Since the statement of the theorem depends only on the equivalence class of $\langle\phi\rangle$, we may assume that ϕ is effective and normalized at P. Suppose $Q \neq P$. Then $\phi(P) \neq \phi(Q)$ if and only if there exists a hyperplane of $\mathbb{P}^n$ containing $\phi(P)$ but not $\phi(Q)$, which is equivalent to $L_\phi(P) \neq L_\phi(Q)$. Since both $L_\phi(P)$ and $L_\phi(Q)$ have codimension 1, we have shown that $\operatorname{codim} L_\phi(P+Q) = 2$ if and only if $\phi(P) \neq \phi(Q)$.

By (4.3.8), we see that $\mathcal{O}_a$ contains a local parameter at P if and only if $L_\phi(2P) \subsetneq L_\phi(P)$, which is equivalent to $\operatorname{codim} L_\phi(2P) = 2$. The result now follows from (4.3.3). □

We now specialize to the case $\langle\phi\rangle = L(D)$ for some divisor D, where we write $\phi = \phi_D$. We implicitly assume that $\dim L(D) \geq 2$ in order to get a map to $\mathbb{P}^n$ for some $n \geq 1$. From the definitions we have $\nu_P(\phi_D) \leq \nu_P(D)$ for all $P \in \mathbb{P}_K$, whence $[\phi_D] \leq D$. We get equality precisely when there is a function $f \in L(D)$ with $\nu_P(f) = -\nu_P(D)$.

Lemma 4.3.14. *Suppose that K has genus $g > 0$, and $D \in \operatorname{Div}(K)$ such that either $\deg D \geq 2g$ or D is canonical. Then $D = [\phi_D]$.*

Proof. The condition we need is that $\dim L(D) > \dim L(D-P)$ for all P. This is immediate from Riemann–Roch when $D-P$ is nonspecial, and in particular when $\deg D \geq 2g$. If D is canonical, then Riemann–Roch yields

$$\dim L(D-P) = 2g-3-g+1+\dim L(P).$$

Since $g > 0$, it follows that $\dim L(P) = 1$, and we have $\dim L(D-P) = g-1 < \dim L(D)$. □

When D is canonical, we call ϕ_D a *canonical* map, or just *the* canonical map if the choice of a particular representative of the equivalence class of ϕ_D doesn't matter. Note that the canonical map is not defined unless K has genus $g \geq 2$.

Theorem 4.3.15. *Suppose that K has genus g and $D \in \mathrm{Div}(K)$. If $\deg D \geq 2g+1$, then ϕ_D is an embedding. If D is canonical, then ϕ_D is an embedding unless K contains a rational subfield $k(x)$ with $|K:k(x)| \leq 2$.*

Proof. Put $\phi := \phi_D$. We have $L_\phi(P+Q) = L(D-P-Q)$ for all $P, Q \in \mathbb{P}_K$ by (4.3.14). If $\deg(D-P-Q) \geq 2g-1$, then $D-P-Q$ is nonspecial. Thus $\operatorname{codim} L_\phi(P+Q) = 2$ by Riemann–Roch, and the result follows from (4.3.13). If D is canonical, Riemann–Roch yields

$$\dim L(D-P-Q) = 2g-4-g+1+\dim L(P+Q) = g-3+\dim L(P+Q).$$

Since $\dim L(D) = g$, (4.3.13) implies that ϕ is an embedding unless $\dim L(P+Q) > 1$. But if there is a nonconstant function $x \in L(P+Q)$, its pole divisor has degree at most two, whence $|K:k(x)| \leq 2$. □

We note that when K contains a rational subfield of index at most 2 the canonical map is not an embedding. Indeed, the canonical map is not defined unless K has genus $g \geq 2$, in which case $|K:k(x)| = 2$, and by (3.6.4) and (3.6.10) we have

$$\Omega_K(0) = \langle \omega, \phi_1(x)\omega, \ldots, \phi_{g-1}(x)\omega \rangle,$$

where ω is a regular differential form and the $\phi_i(x)$ all lie in a (uniquely determined) rational subfield $k(x)$. This means that

$$L([\omega]) = \langle 1, \phi_1(x), \ldots, \phi_{g-1}(x) \rangle,$$

and therefore the image of the canonical map is $\mathbb{P}^1$.

We obtain from (4.3.15) the nonsingularity of the standard plane model for an elliptic curve (2.3.1).

Corollary 4.3.16. *Suppose $g_K = 1$. Then the map ϕ to $\mathbb{P}^2$ of degree three with $\langle \phi \rangle = L(3P)$ for any point P is an embedding.* □

4.4 Weierstrass Points

Following [19], we now turn to a more detailed analysis of projective maps. Let K/k be a function field, let $\phi = (\phi_0, \ldots, \phi_n)$ be a projective map, and let $\tau : K \to K$

be a k-embedding of K into K. Since $|K : k(\tau(x))|$ is finite for any $x \in K \setminus k$, so is $|K : \tau(K)|$. We can therefore define $\deg \tau := |K : \tau(K)|$. Then we have

Lemma 4.4.1. *Denote by* $\tau(\phi)$ *the projective map* $(\tau(\phi_0), \dots, \tau(\phi_n)) : \mathbb{P}_K \to \mathbb{P}^n$. *Then* $\deg \tau(\phi) = \deg \tau \deg \phi$.

Proof. Because they lie in $\tau(K)$, the coordinate functions $(\tau(\phi_0), \dots, \tau(\phi_n))$ define a map $\phi' : \tau(K) \to \mathbb{P}^n$. Since $\tau : K \to \tau(K)$ is an isomorphism, every point of $\tau(K)$ is of the form $\tau(P)$ for some point P of K. In particular, ϕ and ϕ' have the same image in $\mathbb{P}^n$, and therefore $\deg \phi' = \deg \phi$. The result now follows from (4.2.5). □

On first reading, it may be advisable to restrict attention to the important special case $\tau = 1$. In later applications we will want to take τ to be the Frobenius map in characteristic p composed with an automorphism of K.

For any separating variable $s \in K$, let D_s be the Hasse derivative with respect to s given by (1.3.11), and consider the matrix $H = H(\phi, s, \tau) = (h_{ij})$ for $0 \le i \le n$, $j = -1, 0, 1, \dots$, where

$$h_{ij} := \begin{cases} \tau(\phi_i) & \text{if } j = -1, \\ D_s^{(j)}(\phi_i) & \text{for } j = 0, 1, \dots \end{cases} \quad (0 < i \le n).$$

Let $H^{(j)}$ denote the column of H whose i^{th} entry is h_{ij}. Thus, $H^{(-1)}$ is the leading column, and its ith entry is $\tau(\phi_i)$. We are interested in those indices j for which $H^{(j)}$ is not a K-linear combination of lower numbered columns. There are at most $n+1$ such indices since H has $n+1$ rows. Since not all of the ϕ_i are zero, the first such index is always -1. Denote the remaining indices by $j_1, j_2, \dots, j_m$. Note that if $\tau = 1$, then $j_1 > 0$. Conversely, suppose that the first two columns of H are linearly dependent. Then $\tau(\phi) = y\phi$ for some $y \in K$, and we see that $y = \tau(\phi_0)/\phi_0$. Consider the equivalent map $\phi' := \phi_0^{-1}\phi$. Then $\tau(\phi_i') = \phi_i'$ for all i, and we have

Lemma 4.4.2. *With the above notation, if* ϕ *is birational then* $j_1 > 0$ *if and only if* $\tau = 1$. □

If $m < n$, define $j_l = j_{l-1} + 1$ for $m < l \le n$. We will prove that $m = n$ (see (4.4.7)), but in any case, the following property is immediate:

Lemma 4.4.3. *With the above notation, if* $1 \le l \le n$ *and* $j < j_l$, *then* $H^{(j)}$ *is a* K*-linear combination of* $H^{(-1)}, H^{(j_1)}, \dots, H^{(j_{l-1})}$. □

Define $J_s(\phi, \tau) = (j_1, \dots, j_n)$. We call the j_i the τ*-orders* of the map ϕ, or just the orders of ϕ when $\tau = 1$, and we write $J_s(\phi) := J_s(\phi, 1)$. We will show shortly that these indices depend only on the subspace $\langle \phi \rangle$, are independent of s, and are also invariant when ϕ is replaced by $y\phi$ for any $y \in K$. We might expect that the orders of ϕ are just $(1, \dots, n)$ and indeed, we will prove this when the characteristic is zero. In positive characteristic, however, the situation is more complicated.

For any sequence $J = j_1, j_2, \dots$ of nonnegative integers, let H^J be the submatrix of H whose first column is $H^{(-1)}$ and whose $(l+1)$st column is $H^{(j_l)}$ and define

(4.4.4) $$w_s(\phi,\tau) := \det H^{J_s(\phi,\tau)}.$$

We call $w_s(\phi) := w_s(\phi,1)$ the *Wronskian* of ϕ with respect to s.

Lemma 4.4.5. *With the above notation, we have the following.*

1. *If $\phi'_i := \sum_j a_{ij}\phi_j$ $(0 \le i \le n)$, where $a_{ij} \in k$ and $A := (a_{ij})$ is nonsingular, then $J_s(\phi',\tau) = J_s(\phi,\tau)$ and $w_s(\phi',\tau) = \det(A) w_s(\phi,\tau)$.*
2. *For any nonzero function $y \in K$, $J_s(y\phi,\tau) = J_s(\phi,\tau)$ and $w_s(y\phi,\tau) = \tau(y)y^n w_s(\phi,\tau)$.*
3. *For any separating variable $t \in K$, $J_t(\phi,\tau) = J_s(\phi,\tau)$ and*
$$w_t(\phi,\tau) = (ds/dt)^{j_1+\cdots+j_n} w_s(\phi,\tau).$$

Proof. 1) If $H' := H(\phi',s,\tau)$, then $H' = AH$ because the Hasse derivatives and the map τ are all k-linear.

To prove 2), put $\tilde{H} := H(y\phi,s,\tau)$. From the product rule we have

$$\tilde{H}^{(-1)} = \tau(y)H^{(-1)},$$

(∗) $$\tilde{H}^{(j)} = yH^{(j)} + \sum_{k=1}^{j} D_s^{(k)}(y)H^{(j-k)} \quad (j \ge 0).$$

In particular, the K-subspaces spanned by the first j columns of H and $\tilde{H}$ coincide for all j, and we have $J_s(\phi,\tau) = J_s(y\phi,\tau)$. Furthermore, the definition of the j_l and (∗) imply that there is an upper triangular matrix U with entries in K and diagonal entries $(\tau(y), y, y \dots, y)$ such that

(∗∗) $$\tilde{H}^{J_s(\phi,\tau)} = H^{J_s(\phi,\tau)}U,$$

proving 2).

The proof of 3) is similar, with the chain rule replacing the product rule. Here we put $\tilde{H} := H(\phi,t,\tau)$, and (1.3.14) yields functions d_k $(1 \le k < j)$ such that

$$\tilde{H}^{(-1)} = H^{(-1)},$$

$$\tilde{H}^{(j)} = \left(\frac{ds}{dt}\right)^j H^{(j)} + \sum_{k=1}^{j-1} d_k H^{(k)} \quad (j \ge 0).$$

Again, the K-subspaces spanned by the first j columns of H and $\tilde{H}$ coincide, proving that $J_s(\phi,\tau) = J_t(\phi,\tau)$. Moreover, (∗∗) holds again, where now U is upper triangular with $u_{00} = 1$ and $u_{ll} = (ds/dt)^{j_l}$ $(1 \le l \le n)$, and 3) follows. □

We therefore write $J(\phi,\tau) := J_s(\phi,\tau)$ for any separating variable s, and we put

$$j(\phi,\tau) := j_1 + \cdots + j_n.$$

We can now show that $w_s(\phi,\tau) \ne 0$ by evaluation at an appropriate point.

Lemma 4.4.6. *Suppose that $\phi = (\phi_0, \ldots, \phi_n)$ is a projective map, $P \in \mathbb{P}_K$, t is a local parameter at P, and the ϕ_i are defined at P and linearly independent over k. Then the matrix $h_{ij}(P) := D_t^{(j)}(\phi_i)(P)$ $(0 \le i \le n,\ 0 \le j)$ has k-rank $n+1$.*

Proof. We use (2.5.13) to conclude:

1. If all the ϕ are defined at P, so are all the Hasse derivatives, so the statement of the lemma makes sense.

2. If $\alpha_0, \ldots, \alpha_n \in k$, then $\sum_{i=0}^n \alpha_i h_{ij}(P)$ is the coefficient of t^j in the Laurent expansion of $\sum_i \alpha_i \phi_i$ at P.

Since every nonzero function has a nonzero Laurent expansion at P, it follows that a dependence relation on the rows of $h_{ij}(P)$ yields the same dependence relation on the ϕ_i. □

Corollary 4.4.7. *With notation as in* (4.4.4), *if $w_s(\phi, \tau) = 0$, then the ϕ_i are linearly dependent over k.*

Proof. Choose any $P \in \mathbb{P}_K$. Using (4.4.5) to replace ϕ by an equivalent map if necessary, we may assume that s is a local parameter at P and that ϕ is normalized at P. By definition, $w_s(\phi, \tau) = 0$ precisely when the K-rank of the matrix $H = H(\phi, s, \tau)$ is less than $n+1$, in which case there are functions $x_0, \ldots, x_n$, not all zero, such that

$$\sum_{i=0}^{n} x_i D_s^{(j)}(\phi_i) = 0 \quad (j = 0, 1, \ldots).$$

Carefully clearing denominators, we may assume that the x_i are defined at P and are not all zero there. But now evaluating the dependence relation at P, the result follows from (4.4.6). □

Since our standard assumption is that the ϕ_i are linearly independent, we always have $w_s(\phi, \tau) \ne 0$. In fact, the sequence $J(\phi, \tau)$ is characterized as the mimimal sequence of indices for which the corresponding determinant does not vanish:

Corollary 4.4.8. *Suppose $j_1' < \cdots < j_n'$ is a strictly increasing sequence of nonnegative integers, $s \in K$ is a separating variable, and $\det D_s^{(j_l')}(\phi_i) \ne 0$. If $J(\phi, \tau) = (j_1, \ldots, j_n)$, then $j_l \le j_l'$ for all l.*

Proof. This is now immediate from (4.4.3) because for every l, the nonvanishing of the determinant guarantees that the K-rank of $H^{(j_1', \ldots, j_l')}$ must be $l+1$. □

It is also now clear that there is a very close connection between $J(\phi)$ and $J(\phi, \tau)$ for any $\tau \ne 1$. Namely, let s be a separating variable, and put $J(\phi) =: (j_1, \ldots, j_n)$, $H := H(\phi, s, 1)$, and $\tilde{H} := H(\phi, s, \tau)$. Assume that ϕ is birational. Since $H^{J(\phi)}$ is nonsingular, there is, by (4.4.2), some smallest index $m \ge 1$ such that

$$\tilde{H}^{(-1)} \in \langle H^{(-1)}, H^{(j_1)}, \ldots, H^{(j_m)} \rangle.$$

It follows that for $j \geq j_m$,

$$\langle \tilde{H}^{(-1)}, \tilde{H}^{(0)}, \ldots, \tilde{H}^{(j)} \rangle = \langle H^{(-1)}, H^{(0)}, \ldots, H^{(j)} \rangle.$$

Now using the definitions, we have

Lemma 4.4.9. *If $J(\phi) = (j_1, \ldots, j_n)$, $\tau \neq 1$, and ϕ is birational, then there exists an integer $m \geq 1$ such that*

$$J(\phi, \tau) = (0, j_1, \ldots, j_{m-1}, j_{m+1}, \ldots, j_n). \quad \square$$

More importantly, the nonvanishing of $w_s(\phi, \tau)$ and the transformation rules of (4.4.5) allow us to define an invariant divisor as follows. Let s be any separating variable, and put

$$W_s(\phi, \tau) := [w_s(\phi, \tau)] + [\tau(\phi)] + n[\phi] + j(\phi, \tau)[ds].$$

Corollary 4.4.10. *The divisor $W_s(\phi, \tau)$ is independent of s and depends only on τ and the equivalence class of ϕ.*

Proof. Let $a_{ij} \in k$, $y \in K$, and let t be any separating variable. Put $\phi_i' := \sum_j a_{ij}\phi_j$ and let $\phi' := (\phi_0', \ldots, \phi_n')$. From (4.4.5) we have

$$\begin{aligned} w_t(y\phi', \tau) &= \tau(y)y^n w_t(\phi', \tau) = \det(a_{ij})\tau(y)y^n w_t(\phi, \tau) \\ &= \det(a_{ij})\tau(y)y^n (ds/dt)^{j(\phi,\tau)} w_s(\phi, \tau). \end{aligned}$$

Using (4.4.1), we obtain

$$\begin{aligned} W_t(y\phi', \tau) &= [w_t(y\phi', \tau)] + [\tau(y\phi')] + n[y\phi'] + j(\phi, \tau)[dt] \\ &= [\tau(y)] + n[y] + j(\phi, \tau)([ds] - [dt]) + [w_s(\phi, \tau)] - [\tau(y)] + [\tau(\phi')] \\ &\quad - n[y] + n[\phi'] + j(\phi, \tau)[dt] \\ &= [w_s(\phi, \tau)] + [\tau(\phi)] + n[\phi] + j(\phi, \tau)[ds] \\ &= W_s(\phi, \tau). \quad \square \end{aligned}$$

We therefore write $W(\phi, \tau) := W_s(\phi, \tau)$ for any separating variable s (and choice of equivalent map ϕ), and we put $W(\phi) := W(\phi, 1)$. We will call $W(\phi, \tau)$ the *Weierstrass divisor* of ϕ with respect to τ, or just the Weierstrass divisor of ϕ when $\tau = 1$.

The invariance of $W(\phi, \tau)$ is quite powerful. Namely, choose $P \in \mathbb{P}_K$, and recall from section 3.5 that P^τ is the valuation ideal of the discrete valuation $v_P \circ \tau$ of K. We may assume, by a proper choice of coordinates, that ϕ is effective and that $\phi(P)$ and $\phi(P^\tau)$ are both finite. Thus $v_P(\phi) = v_P(\tau(\phi)) = 0$. Let t be a local parameter at P. Then we have

$$W(\phi, \tau) = [w_t(\phi, \tau)] + [\tau(\phi)] + n[\phi] + j(\phi, \tau)[dt],$$

and therefore $v_P(W(\phi, \tau)) = v_P(w_t(\phi, \tau))$. By (2.5.13) the Hasse deriviatives of the coordinate maps ϕ_i are all defined at P, so that $v_P(w_t(\phi, \tau)) \geq 0$. Moreover, (4.4.1) yields the following formula for the degree of $W(\phi, \tau)$:

Theorem 4.4.11. *The divisor $W(\phi,\tau)$ is nonnegative and*

$$\deg W(\phi,\tau) = (\deg\tau + n)\deg\phi + (2g_K - 2)j(\phi,\tau). \quad \square$$

To illustrate the foregoing with an example, choose any $x \in K \setminus kK^p$ and let $\phi := (1,x)$. Taking any separating variable t, the matrix of Hasse derivatives with respect to t is

$$\begin{bmatrix} 1 & 0 & \dots \\ x & dx/dt & \dots \end{bmatrix}$$

Since $dx \neq 0$ by (2.4.6), $J(\phi) = (1)$ and $w_t(\phi) = dx/dt$. Since $[\phi] = [x]_\infty$, we have

$$W(\phi) = [dx/dt] + 2[x]_\infty + [dt] = [dx] + 2[x]_\infty.$$

For a more interesting example, let x be a separating variable and let y be a primitive element for $K/k(x)$ so that $K = k(x,y)$ and $\phi := (1,x,y)$ is a plane model. Now the matrix of Hasse derivatives with respect to x is

$$\begin{bmatrix} 1 & 0 & 0 & \dots \\ x & 1 & 0 & \dots \\ y & dy/dx & D_x^{(2)}(y) & \dots \end{bmatrix}.$$

Provided that $D_x^{(2)}(y)$ does not vanish identically,[2] we have $J(\phi) = (1,2)$, $w_x(\phi) = D_x^{(2)}(y)$, and

$$W(\phi) = [D_x^{(2)}(y)] + 3[\phi] + 3[dx].$$

The points P in the support of $W(\phi,\tau)$ are called the *Weierstrass points* of ϕ with respect to τ, or just the Weierstrass points of ϕ when $\tau = 1$. When $\tau = 1$ and ϕ is the canonical map, they are simply called the Weierstrass points of K, and we put $W(K) := W(\phi)$. Points that are not Weierstrass points are called *ordinary points*. Note that for any ϕ and τ, almost all points are ordinary.

Weierstrass points are of geometric interest, particularly in the case $\tau = 1$, which we now investigate in more detail. Suppose, then, that ϕ is normalized at $P \in \mathbb{P}_K$, that t is a local parameter at P, and that $H(P) := H(\phi,t,1)(P)$ is the matrix with entries in k obtained by evaluating the entries of $H(\phi,t,1)$ at P. As before, let $j'_1 < \dots < j'_n$ be the indices for which the column $H^{(j'_l)}(P)$ is not a k-linear combination of lower-numbered columns, and define $J(\phi)(P) = (j'_1,\dots,j'_n)$. We call the j'_l the *orders of ϕ at P*. Note that the first two columns of $H(P)$ are equal because $\tau = 1$, and they are nonzero because ϕ is normalized at P, so $j'_1 > 0$.

Let $J := (j_1,\dots,j_n)$ be the orders of ϕ, and put $J' := J(\phi)(P)$. If we had $j'_l < j_l$ for some l, the columns of $H^{J'}$ would satisfy a K-dependence relation. Carefully clearing denominators and evaluating at P, we would obtain a nontrivial k-dependence relation on the columns of $H^{J'}(P)$, contrary to the definition of the j'_l. Hence we have $j_l \le j'_l$ for all l. Moreover, if ϕ is normalized at P and t is a

[2]Unfortunately, $D_x^{(2)}(y)$ can vanish identically in characteristic p. See Exercise 4.5.

local parameter at t, it is clear that

$$w_t(\phi)(P) = \det H^J(P),$$

and it follows that $w_t(\phi)$ vanishes at P precisely when $j_l < j'_l$ for some l. We have proved

Lemma 4.4.12. *With the above notation, we have $j_l \leq j'_l$ for all l. Moreover, the following conditions are equivalent:*

1. $j'_l = j_l$ *for all* l.
2. $w_t(\phi)(P) \neq 0$.
3. $\nu_P(W(\phi)) = 0$. □

The orders of ϕ at P have an important geometric intrepretation. Namely, if we row-reduce the matrix $H(P)$ of Hasse derivatives at P and use k-linearity, we obtain a basis $(\phi'_0, \ldots, \phi'_n)$ for $\langle\phi\rangle$ such that if t is a local parameter at P, we have

$$D_t^{(j)}(\phi'_l)(P) = 0 \quad (0 \leq j < j'_l),$$
$$D_t^{(j'_l)}(\phi'_l)(P) = 1.$$

By (2.5.14) we have

Theorem 4.4.13. *Let ϕ be a projective map normalized at P, let t be a local parameter at P, and let $j'_1, \ldots, j'_n$ be the orders of ϕ at P. Define $j'_0 := 0$. Then there exists a basis $(\phi'_0, \ldots, \phi'_n)$ for $\langle\phi\rangle$ such that*

$$\phi'_l = t^{j'_l} + \sum_{j=j'_l+1}^{\infty} c_{l_j} t^j \quad (0 \leq l \leq n),$$

where $c_{l_j} \in k$. In particular, j is an order of ϕ at P if and only if there exists a function $f \in \langle\phi\rangle$ with $\nu_P(f) = j$. □

Recall the notation $L_\phi(D) := \langle\phi\rangle \cap L([\phi] - D)$ and the filtration (4.3.12) from the previous section. Applying (4.4.13), we see that the orders of ϕ at P define the distinct subspaces of this filtration; namely, we have

$$L_\phi(j'_l P) = \langle\phi'_l, \ldots, \phi'_n\rangle \quad (0 \leq l \leq n).$$

Corollary 4.4.14. *If ϕ is nonsingular at P, then the osculating filtration at P is*

$$\langle\phi\rangle = L_\phi(0) \supsetneq L_\phi(P) \supsetneq L_\phi(2P) \supseteq \cdots.$$

In particular, $j'_1 = 1$.

Proof. This follows from (4.3.13), since $\operatorname{codim} L_\phi(2P) = 2$. □

Geometrically, each linear subspace $L \subseteq \mathbb{P}^n$ of dimension l is the zero set of some set of $n - l$ independent linear forms $\mathscr{L}$. If we choose a hyperplane at infinity defined by some linear form ℓ_∞ that does not vanish at P and put $L^* := \langle\phi^*(\ell/\ell_\infty) \mid \ell \in \mathscr{L}\rangle$, we get a bijection $L^* \leftrightarrow L$ between $(n-l)$-dimensional subspaces of $\langle\phi\rangle$ and l-dimensional subspaces of $\mathbb{P}^n$. Under this correspondence, the subspaces of $\mathbb{P}^n$ corresponding to the $L_\phi(j'_l P)$ are called *osculating planes*. In particular, the hyperplane corresponding to $L_\phi(j'_n P)$ is called the *osculating hyperplane at P*. See Exercise 4.6

When ϕ is the canonical map, the osculating filtration at P is particularly interesting, because by (4.3.14), we have $\langle\phi\rangle = L([\phi]) \supseteq L([\phi] - jP)$, whence Riemann–Roch yields

$$\dim L(jP) = j - g + 1 + \dim L_\phi(jP)$$

for any integer j. Hence

$$\dim L((j+1)P)/L(jP) = 1 - \dim L_\phi(jP)/L_\phi((j+1)P).$$

Since j is a canonical order at P precisely when $L_\phi(jP) \supsetneq L_\phi((j+1)P)$, we see that j is a canonical order at P if and only if there is no function in K with a pole of order exactly $j+1$ at P and no other poles. When $L((j+1)P) = L(jP)$ we say that $j+1$ is a *gap number* at P. So we have

Corollary 4.4.15. *The positive integer j is an order of the canonical map at P if and only if $j+1$ is a gap number at P.* □

The positive gap numbers have interesting properties. Call the integer j a *pole number* at P if there exists a function f with a pole of order exactly j at P and no other poles.

Lemma 4.4.16. *If n is a gap number, then at least half of the positive integers less than n are also gap numbers.*

Proof. Clearly, if j and k are pole numbers, then taking the product of the corresponding two functions shows that $j+k$ is also a pole number. It follows that if n is a gap number and $m < n$ is a pole number, then $n-m$ must be a gap number. Thus, there are at least as many gap numbers less than n as pole numbers. □

Corollary 4.4.17. *Let $j'_1, \ldots, j'_{g-1}$ be the orders of the canonical map at some point $P \subset \mathbb{P}_K$. Then $j'_l \leq 2l$ for all l.*

Proof. If $n_1, n_2, \ldots, n_l$ are the first l positive gap numbers, then we have $l \geq (n_l - 1)/2$ by (4.4.16). Note that $n_1 = 1$; otherwise K is rational, and there is no canonical map. Thus, using (4.4.15), we see that $j'_{l-1} = n_l - 1$, and the inequality follows. □

Corollary 4.4.18 (Clifford). *For $n \leq 2g-2$ and $P \in \mathbb{P}_K$ we have* $\dim L(nP) \leq 1 + n/2$.

Proof. By Riemann–Roch, $\dim L((2g-2)P) \leq g$. We may therefore assume, by descending induction on n, that the formula holds for $n+1 \leq 2g-2$. If $\dim L((n+1)P) = \dim L(nP) + 1$, the result follows. Otherwise, $n+1$ is a gap number, and the result follows from (4.4.16). □

Returning now to the case of general τ, recall from Section 3.5 that for any $P \in \mathbb{P}_K$ we have defined $P^\tau := \tau^{-1}(P)$, the set-theoretic inverse image of τ. In particular, P is fixed by τ precisely when $\tau(P) \subseteq P$. We will call P a *strong fixed point* of τ if $\tau(P) \subseteq P^2$.

Suppose that ϕ is birational and $\tau \neq 1$. Then $j_1 = 0$ by (4.4.2). At a fixed point P of τ, however, the first two columns of $H(\phi, s, \tau)$ have the same value and thus $v_P(w_s(\phi, \tau)) > 0$ for any s. We therefore have

Lemma 4.4.19. *With the above notation, suppose that ϕ is birational and $\tau \neq 1$. Then every fixed point of τ is a Weierstrass point of ϕ with respect to τ.* □

In particular, the number of fixed points of τ is at most $\deg W(\phi, \tau)$ for any birational map ϕ. However, this bound is rather crude, and can be improved by lower-bounding the multiplicity of each Weierstrass point P dividing $W(\phi, \tau)$. To do this, let $I := (i_1, \ldots, i_n)$ and $J := (j_1, \ldots, j_n)$ be nonnegative integer sequences, and let

$$\binom{J}{I} := \binom{j_r}{i_s}$$

denote the $n \times n$ matrix of binomial coefficients, where $\binom{j}{i} = 0$ if $j < i$, and the binomial coefficients are interpreted as elements of the ground field, i.e., they are reduced modulo p if $\mathrm{char}(K) = p$.

Define

$$t^I := \mathrm{diag}(t^{i_1}, \ldots, t^{i_n}),$$

and write $I \leq J$ if $i_l \leq j_l$ for all l.

Theorem 4.4.20. *Let K/k be a function field and let $\tau : K \to K$ be a k-embedding. Let $P \in \mathbb{P}_K$ be a fixed point of τ, let $\phi : \mathbb{P}_K \to \mathbb{P}^n$ be effective and normalized at P, and let $t \in K$ be a local parameter at P. Let $J := j(\phi)(P) = (j_1, \ldots, j_n)$ be the orders of ϕ at P and let $I := (i_1 < i_2 < \cdots < i_n)$ be an increasing sequence of nonnegative integers. Then*

$$v_P(\det H(\phi, t, \tau)^I) \geq \sum_{l=1}^{n} (j_l - i_l).$$

Moreover, if $i_1 > 0$ or if P is a strong fixed point of τ, then equality holds if and only if

$$\det \binom{J}{I} \neq 0.$$

Proof. By (4.4.13) there is a basis $\phi'_l = \sum_j a_{lj} \phi_j$ for $\langle \phi \rangle$ such that $\phi'_0 = 1$ and

$$\phi'_l = t^{j_l} + t^{j_l+1} v_l \quad (0 < l \leq n),$$

where $v_l \in \mathcal{O}_P$ and $j_l > 0$. Then (2.5.13) yields

$$D_t^{(i_r)}(\phi'_l) = \binom{j_l}{i_r} t^{j_l - i_r} + v_{l_r} t^{j_l - i_r + 1} \quad (0 \leq l \leq n,\ 0 < r \leq m),$$

where $v_{lr} \in \mathscr{O}_P$. Put $H := H(\phi,t,\tau)$, $A := (a_{ij})$, and $m := \sum_l (j_l - i_l)$.

If $i_1 > 0$ then the first row of H^I is $(1,0,0,\dots)$. Expanding the determinant along the first row yields

$$(*) \qquad \det H^I = \det\left[t^J A^{-1}\left(\binom{J}{I} + tV\right)t^{-I}\right] = \det(A)^{-1} t^m \det\left(\binom{J}{I} + tV\right),$$

where $V = (v_{lr})$, and the theorem follows in this case.

If $i_1 = 0$, then the first row of H^I is $(1,1,0,0,\dots)$. Replace the second column by itself minus the first, and again expand along the first row. The only difference from the previous case is that in the first column of the minor, ϕ_l' is replaced by $\phi_l' - \tau(\phi_l')$. If P is a strong fixed point of τ, we have

$$\tau(\phi_l') = t^{j_l'} + t^{j_l'+1} v_l'$$

for some $j_l' > j_l$ and some $v_l' \in \mathscr{O}_P$, and thus $(*)$ still holds with V suitably redefined.

Finally, if $i_1 = 0$ but we don't know that P is a strong fixed point, then we might have $v_P(\phi_l' - \tau(\phi_l')) > j_l$. In this case, an equation similar to $(*)$ will hold, with $\binom{J}{I}$ replaced by another integer matrix which may have a few more zeroes in the first column, but the inequality still follows. □

There are several important corollaries to (4.4.20). Let $p := \operatorname{char}(k)$ in all of these. First, we get a lower bound on the order of the Weierstrass divisor $W(\phi,\tau)$ at a fixed point of τ by taking I above to be the order sequence of ϕ with respect to τ and applying (4.4.10).

Corollary 4.4.21. *If $J := J(\phi,\tau) = j_1,\dots,j_n$ are the τ-orders of ϕ and $J' := J(\phi)(P) = j_1',\dots,j_n'$ are the orders of ϕ at a fixed point P of τ, then*

$$v_P(W(\phi,\tau)) \geq \sum_{l=l}^{n} (j_l' - j_l).$$

Moreover, if $\tau = 1$ or if P is a strong fixed point of τ, then equality holds if and only if

$$\det\binom{J'}{J} \neq 0. \quad \square$$

As a further application of (4.4.20), we can show that the τ-orders of ϕ are "generically" either $(1,2,\dots,n)$ or $(0,1,\dots,n-1)$.

Lemma 4.4.22. *Assume that either* $\operatorname{char}(k) = 0$ *or* $\operatorname{char}(k) > \deg\phi$, *and let $J := (j_1,\dots,j_n)$. If $I := (1,2,\dots,n)$, then*

$$\det\binom{J}{I} = \prod_{l=1}^{n} \frac{j_l}{l!} \prod_{m=1}^{l-1} j_l - j_m.$$

If $I := (0,1,2,\ldots,n-1)$, *then*

$$\det\binom{J}{I} = \prod_{l=1}^{n} \frac{1}{l-1!} \prod_{m=1}^{l-1} j_l - j_m.$$

In particular, if ϕ *is birational then*

$$J(\phi,\tau) = \begin{cases} (1,2,\ldots,n) & \textit{if } \tau = 1, \\ (0,1,\ldots,n-1) & \textit{otherwise.} \end{cases}$$

Proof. Expanding the binomial coefficient, we have

$$\binom{j_l}{m} = \frac{j_l(j_l-1)\cdots(j_l-m+1)}{m!} = \frac{j_l^m + \sum_{i=1}^{m-1} v_{lmi} j_l^{m-i}}{m!},$$

where the v_{lmi} are integers. Thus, we can write

$$\binom{J}{I} = D_1 V_1 D_2^{-1},$$

where $D_1 := \operatorname{diag}(j_1, j_2, \ldots, j_n)$, $D_2 := \operatorname{diag}(1/1!, 1/2!, \ldots, 1/n!)$, and V_1 is an integer matrix that is evidently equivalent by unimodular column operations to the Vandermonde matrix $V = V(j_1,\ldots,j_n)$ whose (l,m)-entry is j_l^{m-1}. Since $\det V = \prod_{m<l} j_l - j_m$, the first formula follows, and the proof of the second is similar.

Now choose any point $P \in \mathbb{P}_K$, let $J(\phi)(P) = j_1,\ldots,j_n$, and let $I = (1,2,\ldots,n)$ if $\tau = 1$ and $(0,1,\ldots,n-1)$ otherwise. Clearly, we have $J(\phi,\tau) \geq I$ by minimality of I. Since j_n is the order of a linear functional on $\mathbb{P}^n$ at the point $\phi(P)$ on the image of ϕ, we have $j_n \leq \deg\phi$. Thus, the above formulas and (4.4.20) imply that H^I is nonsingular, whence $J(\phi,\tau) \leq I$, and we therefore have equality. □

We call the projective map ϕ *classical* when $J(\phi) = 1,\ldots,n$. To construct a nonclassical map to $\mathbb{P}^2$, let $\phi := (1, x, y^p)$, where x is a separating variable, $K = k(x,y)$, and $\operatorname{char}(k) = p > 2$. Then $K/k(x,y^p)$ is both separable and purely inseparable, so $K = k(x,y^p)$. We may choose $y \notin K^p$ so that y is also a separating variable by (2.4.6). In particular, $dy/dx \neq 0$. Moreover, $D_y^{(j)}(y^p) = 0$ for $1 \leq j < p$. Now (1.3.14) implies that $D_x^{(j)}(y^p) = 0$ for $1 \leq j < p$, and that $D_x^{(p)}(y^p) = (dy/dx)^p$. The matrix of Hasse derivatives with respect to x is therefore

$$\begin{bmatrix} 1 & 0 & \ldots & 0 & \ldots \\ x & 1 & \ldots & 0 & \ldots \\ y^p & 0 & \ldots & (dy/dx)^p & \ldots \end{bmatrix}.$$

It follows that the order sequence of ϕ is $(0,1,p)$ and $W_x(\phi) = (dy/dx)^p$. However, this particular ϕ is singular. See Exercise 4.5 for a more interesting nonsingular example.

Finally, we obtain a powerful bound on the number of strong fixed points of τ due to Stöhr–Voloch [19]. The key lemma is

Lemma 4.4.23. *With the above notation, let P be a strong fixed point of τ, let $J(\phi,\tau) =: j_1, j_2, \ldots, j_n$, and let $J' := j'_1, \ldots, j'_n$ be the orders of ϕ at P. Assume that $\tau(\phi)$ is not a K-multiple of ϕ. Then $j_l \le j'_l - j'_1$ for $1 \le l \le n$.*

Proof. For any integer sequence $I = i_1, \ldots, i_n$, define $I^- := i_2, \ldots, i_n$ and $I^+ := 0, i_1, \ldots, i_n$. Put $\tilde{J} := j'_2 - j'_1, \ldots, j'_n - j'_1$. Since we are assuming that $j_1 = 0$, it suffices to show that $J(\phi,\tau)^- \le \tilde{J}$, or equivalently, that $J(\phi,\tau) \le \tilde{J}^+$.

Consider the map $\psi := (1, t^{j'_2 - j'_1}, \ldots, t^{j'_n - j'_1}) : \mathbb{P}_{k(t)} \to \mathbb{P}^{n-1}$. Since $\tilde{J}$ is the order sequence of ψ at (t), we have $J(\psi) \le \tilde{J}$. It therefore suffices to show that $J(\phi,\tau)^- \le J(\psi)$ or equivalently that $J(\phi,\tau) \le J(\psi)^+$.

Multiplying by t^{j_1} yields the equivalent effective map $\psi' := (t^{j'_1}, t^{j'_2}, \ldots, t^{j'_n})$ whose order sequence at (t) is J', and by (4.4.5) we have $J(\psi) = J(\psi')$. Hence, the matrix

$$H(\psi', t, 1)^{J(\psi)} = t^{J'} \binom{J'}{J(\psi)^+} t^{-J(\psi)^+}$$

is nonsingular. In particular, we get

$$\det \binom{J'}{J(\psi)^+} \neq 0.$$

Now (4.4.20) implies that $H(\phi,t,\tau)^{J(\psi)^+}$ is nonsingular, and thus $J(\phi,\tau) \le J(\psi)^+$ by definition of $J(\phi,\tau)$. □

Theorem 4.4.24 (Stöhr–Voloch). *Let K/k be a function field of genus g, let $\tau : K \to K$ be a k-embedding, and let $\phi : \mathbb{P}_K \to \mathbb{P}^n$ be a projective map. Assume that $\tau(\phi)$ is not a K-multiple of ϕ. Then the number of strong fixed points of τ is at most*

$$\frac{1}{n} \deg W(\phi,\tau) = (1 + \frac{\deg \tau}{n}) \deg \phi + \frac{2g-2}{n} j(\phi,\tau).$$

Proof. Continuing the notation of the previous lemma, we have shown that $j_l \le j'_l - j'_1$ for $1 \le l \le n$. We conclude from (4.4.20) that

$$v_P(W(\phi,\tau)) \ge n j'_1 \ge n,$$

because $j'_1 \ge 1$ by hypothesis. Since every strong fixed point of τ divides $W(\phi,\tau)$ by (4.4.19), the theorem follows from (4.4.11). □

We can specialize the above result to the case $\phi = \phi_D$, the projective map determined by $L(D)$ for some nonspecial divisor D. In particular, we get

Corollary 4.4.25. *Let K/k be a function field of genus g and let $\tau : K \to K$ be a k-embedding. Then for any integer $n > g$, the number of strong fixed points of τ is at most*

$$1 + \deg \tau + \left(n + \frac{\deg \tau}{n}\right) g + \frac{2g^2(g-1)}{n}.$$

Proof. We may as well assume that there is at least one strong fixed point, P, or there is nothing to prove. Let $d := n+g \geq 2g+1$. Then $\dim L(dP) = n+1$ by Riemann–Roch (2.2.9). Let $\phi' = (1, \phi'_1, \ldots, \phi'_n)$ be a basis for $L(dP)$ adapted to the filtration

$$L(0) \subseteq L(P) \subseteq \cdots \subseteq L(dP).$$

Note that $\deg \phi' = d$ by (4.3.14), and that ϕ' is non-singular by (4.3.15). In particular, ϕ' is birational and thus $\tau(\phi)$ is not a K-multiple of ϕ by (4.4.2). Moreover, $v_P(\phi'_{n-i}) \geq -d+i$ for all i, with equality for $i < d-2g$ by (2.2.9). It follows that the map $\phi := (\phi'_n)^{-1}\phi'$ is normalized at P and the orders of ϕ at P are

$$1, 2, \ldots, d-2g-1, j'_{d-2g}, \ldots, j'_n = d.$$

All we know about the last $g-1$ orders j'_i is that they are distinct integers in the closed interval $[i, d]$, which yields the bound $j'_i \leq i+g$. Using (4.4.23) we have

$$j(\phi, \tau) \leq \sum_{i=1}^{n} (j'_i - 1) \leq g(g-1) + \frac{n(n-1)}{2}.$$

Substituting this bound into (4.4.24) and simplifying, we see that the number of strong fixed points of τ is at most

$$\begin{aligned} &\left(1 + \frac{\deg \tau}{n}\right) \deg \phi + \frac{2g-2}{n} j(\phi, \tau) \\ &\qquad \leq \left(1 + \frac{\deg \tau}{n}\right)(n+g) + \frac{2g(g-1)^2}{n} + (g-1)(n-1)) \\ &\qquad = 1 + \deg \tau + \left(n + \frac{\deg \tau}{n}\right) g + \frac{2g^2(g-1)}{n}. \qquad \square \end{aligned}$$

We will apply (4.4.25) in Chapter 5 to prove the Riemann hypothesis for curves over finite fields.

We conclude this section with an application to the automorphism group of a function field, by which we mean the group $\mathrm{Gal}(K/k)$ of automorphisms that are the identity on constants. We first prove

Lemma 4.4.26. *Let K be a function field of genus g and let $\sigma \in \mathrm{Gal}(K/k)$. If σ fixes more than $2g+2$ points of $\mathbb{P}_K$ then $\sigma = 1$.*

Proof. By (4.4.19), σ fixes only finitely many points. Let P be a point not fixed by σ. By Riemann–Roch, there is a nonconstant function $x \in L((g+1)P)$. Then $\sigma^{-1}(x) \in L((g+1)P^\sigma)$, so $x - \sigma^{-1}(x)$ and $\sigma(x) - x$ have pole divisors of degree at most $2(g+1)$. Since any fixed point of σ is a zero of $\sigma(x) - x$, the result follows. $\square$

For the remainder of this discussion we will assume that either $\mathrm{char}(k) = 0$ or $\mathrm{char}(k) = p > 2g-2$. Let ϕ be the canonical map and let $W := W(\phi)$. Note that by (4.4.11) and (4.3.14), $\deg W = (g + j(\phi))(2g-2)$. Then (4.4.22) yields $j(\phi) = g(g-1)/2$, whence

$$\deg W = (g-1)g(g+1),$$

and for any point P we have

$$v_P(W) = \sum_{l=0}^{g-1} (j_l(P) - l).$$

The integer $v_P(W)$ is called the *weight* of the Weierstrass point P. From (4.4.17) we get $v_P(W) \leq g(g-1)/2$, and it follows that K has at least $2g+2$ distinct Weierstrass points. We claim that for $g > 1$, the subgroup of $\mathrm{Gal}(K/k)$ that fixes all the Weierstrass points has order at most 2, and in particular, that $\mathrm{Gal}(K/k)$ is a finite group. If the number of Weierstrass points is greater than $2g+2$ this follows from (4.4.26).

If K has exactly $2g+2$ Weierstrass points, then equality holds in (4.4.26) for all l and all Weierstrass points P. Choose a Weierstrass point and call it P_∞. We have $j_1(P_\infty) = 2$, whence 2 is not a gap number by (4.4.16). We therefore have a nonconstant function $x \in L(2P_\infty)$, so K is hyperelliptic. Since $\mathrm{char}(k) > 2$, (3.6.2) and the remark following it yield $K = k(x,y)$ with $y^2 = f(x)$, where $f(x)$ is a square-free polynomial of degree $2g+1$. Moreover, if the roots of f are $\{a_1, \ldots, a_{2g+1}\}$, the ramified points of $K/k(x)$ are the unique points P_i dividing $x - a_i$ together with P_∞. Thus, $(x-a_i)^{-1} \in L(2P_i)$ for all i, and it follows that the Weierstrass points are $\{P_\infty, P_1, \ldots, P_{2g+1}\}$. Any automorphism σ acts on $k(x)$, the unique rational subfield of K of index 2. If σ fixes P_∞ then σ acts on $k[x]$, and it is easy to see that $\sigma(x) = ax + b$ for some $a, b \in k$. If σ also fixes $x - a_i$ for all i, then $a = b = 1$ since $2g+1 > 2$. So in this case, the group of automorphisms that fixes all the Weierstrass points has order 2. We have proved

Theorem 4.4.27. *Let K/k be a function field of genus $g > 1$ where either* $\mathrm{char}(k) = 0$ *or* $\mathrm{char}(k) > 2g-2$. *Then one of the following holds:*

1. *K has more than $2g+2$ Weierstrass points and* $\mathrm{Gal}(K/k)$ *permutes them faithfully.*

2. *K has exactly $2g+2$ Weierstrass points, and the subgroup of* $\mathrm{Gal}(K/k)$ *fixing all these points is* $\mathrm{Gal}(K/k(x))$ *for some $x \in K$ with $|K : k(x)| = 2$.*

In particular, $\mathrm{Gal}(K/k)$ *is a finite group.* □

For an example of a hyperelliptic curve (in characteristic 2 !) with exactly one Weierstrass point, see Exercise 4.3.

4.5 Plane Curves

In this section we apply our results on projective curves to the important special case of plane curves. One reason for the importance of plane curves is that we can always write a function field K as $k(x,y)$ by choosing a separating variable x and a primitive element y for $K/k(x)$. This yields a map $\phi := (1,x,y)$ to $\mathbb{P}^2$ with image V such that $K = k(V)$. Such a curve V is called a *plane model* for K. In general, however, V may have singularities. Indeed, as we will see, there exist curves for which every plane model is singular. For the remainder of this section we will be assuming that ϕ is an effective map to $\mathbb{P}^2$, which, unless otherwise specified, will be written $\phi = (1,x,y)$.

A line L in $\mathbb{P}^2$ is just the set of zeros of a homogeneous linear form $\ell := aX_0 + bX_1 + cX_2$. Since $\mathbf{V}(\ell)$ is uniquely determined by the triple (a,b,c) up to a nonzero scalar multiple, we often abuse notation by writing $L = (a:b:c)$. In this way, the set of lines form another $\mathbb{P}^2$ called the *dual plane*.

Recall that for a homogeneous polynomial $g(X_0,X_1,X_2)$, the set of all points of intersection of $\mathbf{V}(g)$ with V, together with their multiplicities, is given by the divisor $[\phi]_g = [\phi] + [g^*]$, and the intersection multiplicity of V and $\mathbf{V}(g)$ at a point $a \in \mathbb{P}^2$ is given by

$$\mu_a(\mathbf{V}(g)) = \sum_{\phi(P)=a} v_P([\phi]_g).$$

By (4.1.10), a plane curve V is always the zero set of a single irreducible polynomial f. If g is irreducible, then $\mathbf{V}(g)$ is another plane curve, and we might ask whether the intersection multiplicity we have defined is symmetric in f and g. The affirmative answer follows from an important alternative description of the intersection multiplicity, which we now derive. The starting point is

Lemma 4.5.1. *Suppose $S \subseteq R$ are k-algebras, R is an integral domain, and $x \in S$. If R/S and R/Rx are finite-dimensional, then* $\dim_k(R/Rx) = \dim_k(S/Sx)$.

Proof. Consider the inclusion diagram

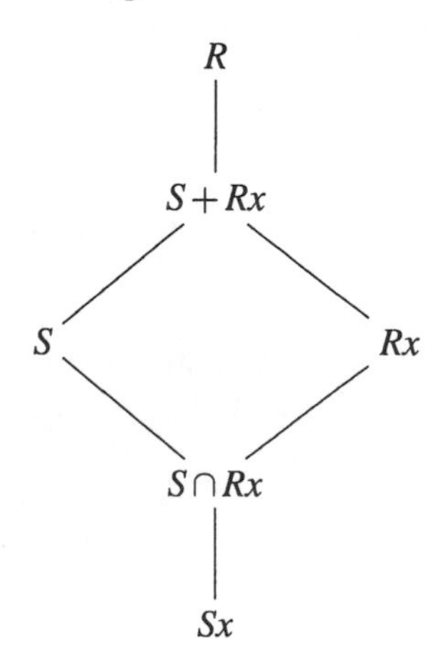

Multiplication by x induces an isomorphism $R/S \simeq Rx/Sx$. It follows that R/Sx is finite-dimensional, and thus

$$\dim_k(R/(S+Rx)) + \dim_k((S+Rx)/S) = \dim_k(Rx/(S\cap Rx)) + \dim_k((S\cap Rx)/Sx).$$

Moreover, we have

$$(S+Rx)/S \simeq Rx/(S\cap Rx), \quad \text{and} \quad (S+Rx)/Rx \simeq S/(S\cap Rx)$$

by the isomorphism theorems. Using these isomorphisms, we obtain

$$\dim_k(R/(S+Rx)) + \dim_k((S+Rx)/Rx) = \dim_k(S/(S\cap Rx)) + \dim_k((S\cap Rx)/Sx),$$

and the result follows. □

Theorem 4.5.2. *Let $V = \mathbf{V}(f)$ be a plane projective curve with irreducible defining polynomial f, let $g = g(X_0, X_1, X_2)$ be any homogeneous polynomial, and let $a \in V \cap \mathbf{V}(g)$. Choosing notation so that $X_0(a) \neq 0$, let M be the image of $\mathbf{I}(a)$ in the polynomial ring $k[X_1, X_2]$ after specializing $X_0 = 1$, let $A := k[X_1, X_2]_M$ be the localization at M, and let I be the ideal of A generated by $f(1, X_1, X_2)$ and $g(1, X_1, X_2)$. Then*

$$\mu_a(\mathbf{V}(g)) = \dim_k A/I. \tag{4.5.3}$$

In particular, if g is irreducible and $\tilde{\mu}_a(\mathbf{V}(f))$ is the intersection multiplicity of $\mathbf{V}(f)$ as defined on the curve $\mathbf{V}(g)$, then $\mu_a(\mathbf{V}(g)) = \tilde{\mu}_a(\mathbf{V}(f))$.

Proof. Letting $g^* := g(1, X_1, X_2)$ restricted to V as usual, we have $A/(f) \simeq \mathcal{O}_a$ and $A/(f, g) \simeq \mathcal{O}_a/\mathcal{O}_a g^*$. Let R_a be the integral closure of $\mathcal{O}_a$ in $K := k(V)$. By (4.3.2) and (1.1.17), we have $\dim_k(R_a/\mathcal{O}_a) < \infty$ and $\dim_k(R_a/R_a g^*) < \infty$. We can therefore apply (4.5.1) to conclude that

$$\dim_k(\mathcal{O}_a/\mathcal{O}_a g^*) = \dim_k(R_a/R_a g^*).$$

Let $\phi : \mathbb{P}_K \to V$ be the natural map, and let $\mathscr{V}$ denote the set of valuations of K corresponding to the points in $\phi^{-1}(a)$. Define $m(v) = v(g^*)$ for $v \in \mathscr{V}$. Then from (1.1.17) we have $R_a = K(\mathscr{V}; 0)$, $R_a g^* = K(\mathscr{V}; m)$, and

$$\dim_k(R_a/R_a g^* = \sum_{v\in\mathscr{V}} m(v) = \sum_{P\in\phi^{-1}(a)} v_P(g^*) = \mu_a(\mathbf{V}(g)). \quad \square$$

What happens when we specialize the above result to the case that g is a linear form? In the first place, if we interchange the roles of f and g, we are applying the theory developed in this section to a line L in the plane. Here, the natural map $(1, x, y)$ degenerates because there is a dependence relation $a + bx + cy = 0$. Choosing notation so that $c \neq 0$, we have $K = k(x, y) = k(x)$. Since x has just one pole and it is simple, $\deg L = 1$. Now (4.2.8) says that $\mathbf{V}(f)$ meets L in $\deg(f)$ points, and we have proved

Corollary 4.5.4. *If $V = \mathbf{V}(f)$ is a plane curve, then* $\deg V = \deg f$. □

We can make this more explicit by looking closely at a point $a \in V \cap L$. Translate coordinates so that $a = (1:0:0)$ and let $\hat{f} := f(1,x,y)$ be the dehomogenized defining polynomial. Then A is the ring of rational functions in x and y whose denominators have a nonzero constant term, and I is the ideal generated by $\hat{f}$ and some linear polynomial $bx+cy$. To understand A/I, first factor out $(bx+cy)$ and put $\bar{A} := A/(bx+cy)$. Then $\bar{A}$ is just the discrete valuation ring of rational functions in one variable whose denominators have a nonzero constant term, and (choosing notation so that $c \neq 0$), the image of I in $\bar{A}$ is generated by the polynomial $\bar{f}(x)$ obtained by substituting $y = -bx/c$ in $\hat{f}$. Write $\bar{f} =: x^e f_0(x)$, where $f_0(0) \neq 0$. Then

$$\dim_k A/I = \dim_k \bar{A}/(\bar{f}) = e.$$

We call e the *order of vanishing of f at a along L*. Then (4.5.2) specializes to

Corollary 4.5.5. *Let $V = \mathbf{V}(f)$ be a plane curve, let $L \subseteq \mathbb{P}^2$ be a line, and let $a \in V \cap L$. Then the intersection multiplicity of L at a equals the order of vanishing of f at a along L.* □

Recall from (4.4.14) that at each point $P \in \mathbb{P}_K$ the osculating filtration at P is

$$\langle \phi \rangle = L_\phi(0) \supsetneq L_\phi(j_1 P) \supsetneq L_\phi(j_2 P) \supsetneq 0,$$

where $\{0, j_1, j_2\}$ are the orders of ϕ at P. This says that all lines through $\phi(P)$ have intersection multiplicity j_1 at P except for the osculating line which has multiplicity j_2. The nonosculating lines at P are called *generic* at P. We say that a line is generic at a point $a \in V$ if it is generic at every point $P \in \phi^{-1}(a)$. When ϕ is nonsingular at P, then generic lines have intersection multiplicity 1 and the osculating line is called the *tangent line*.

Lemma 4.5.6. *Let $\phi = (x_0, x_1, x_2)$ be any plane map $\mathbb{P}_K \to V \subseteq \mathbb{P}^2$, and suppose that ϕ is normalized and nonsingular at P. Let $a_0, a_1, a_2 \in k$. Then the equation of the tangent line to V at $\phi(P)$ is $\sum_i a_i X_i = 0$ if and only if the following two conditions hold:*

$$\sum_i a_i x_i(P) = 0,$$

$$\sum_i a_i \frac{dx_i}{dt}(P) = 0,$$

where t is a local parameter at P.

Proof. Put $\ell := \sum_i a_i x_i$. Then the first condition is equivalent to $v_P(\ell) \geq 1$. When this occurs, expanding ℓ in a power series in t and using (2.5.14) we see that the two conditions together are equivalent to $v_P(\ell) \geq 2$. □

We next make the important observation, which is often taken as the definition, that the partial derivatives of the defining equation give the coordinates of the tangent line at all nonsingular points.

Lemma 4.5.7. *Let $V = \mathbf{V}(f)$ and put $f_i := \partial f/\partial X_i$ for $i = 0,1,2$. If V is nonsingular at a, then the tangent line at a is $(f_0(a) : f_1(a) : f_2(a))$.*

Proof. First of all, if f has degree d, then each f_i has degree $d-1$, so that $(f_0(a) : f_1(a) : f_2(a))$ is a well-defined point of the dual plane. Renumbering the coordinate axes if necessary, we may assume that $X_0(a) \neq 0$. As usual, we have $x := X_1/X_0$, $y := X_2/X_0$, and the identity

$$\hat{f}(x,y) := \frac{f(X_0,X_1,X_2)}{X_0^d} = 0. \tag{4.5.8}$$

Using the fact that f is homogeneous, we verify the identity

$$\sum_i X_i f_i = \deg(f) f,$$

from which it follows that the linear form

$$\ell := \sum_i \frac{f_i(a)}{X_0^{d-1}(a)} X_i$$

vanishes at a, and therefore $\ell^* := \ell/X_0$ vanishes at P. We next check that

$$\begin{aligned} \hat{f}_x &:= \frac{\partial \hat{f}}{\partial x} = \frac{f_1}{X_0^{d-1}}, \\ \hat{f}_y &:= \frac{\partial f}{\partial y} = \frac{f_2}{X_0^{d-1}}. \end{aligned} \tag{4.5.9}$$

This implies that $\ell^* = f_0(a)/X_0^{d-1}(a) + \hat{f}_x(P)x + \hat{f}_y(P)y$. Differentiating, we have $d\ell^* = \hat{f}_x(P)dx + \hat{f}_y(P)dy$. However, differentiating (4.5.8), we have the identity

$$\hat{f}_x dx + \hat{f}_y dy = 0, \tag{4.5.10}$$

which implies that $v_P(d\ell^*) = 0$. The result now follows from (4.5.6). □

Because the partial derivatives are all homogeneous of the same degree, the map $\psi(a) := (f_0(a) : f_1(a) : f_2(a))$ is well-defined at any point in the plane at which at least one partial derivative is nonzero. Moreover, if $K = k(V)$ and $\phi : \mathbb{P}_K \to V$ is the natural map, there is a well-defined projective map from $\mathbb{P}_K$ to the dual plane given by

$$\tilde{\phi} := (f_0(1,x,y), f_1(1,x,y), f_2(1,x,y)), \tag{4.5.11}$$

whose image $\tilde{V}$ is called the *dual curve*. This yields a diagram

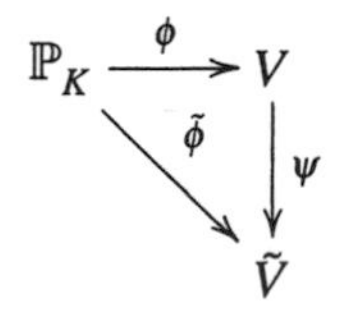

that commutes wherever ψ is defined, namely at all points $P \in \mathbb{P}_K$ at which ϕ is nonsingular. Note, however, that $\tilde{\phi}$ is defined at all points $P \in \mathbb{P}_K$.

Some care needs to be taken here. In characteristic zero one can show that the map $\tilde{\phi}$ is birational, and that when V is nonsingular of degree $d > 2$, $\tilde{V}$ has degree $d(d-1)$ and is in fact singular. In characteristic p, however, the situation is trickier. See Exercise 4.5.

Changing topics, we turn now to a discussion of the adjoint divisor. Given a plane curve $V = \mathbf{V}(f)$ we let $\hat{f} := f(1,x,y)$ be the dehomgenization of the defining polynomial as usual, and we consider the differential form $dx/\hat{f}_y$, where $\hat{f}_y := \partial \hat{f}/\partial y$. Since this form depends on a choice of coordinates in projective space, we first need to understand the nature of the dependence. Let X denote the column vector $(X_0, X_1, X_2)^t$ of coordinates in $\mathbb{P}^2$, and let $A = a_{ij} \in GL(3,k)$ be a nonsingular change of variable, so that $X = AX'$. Pulling back to the curve, we have functions $x' := X_1'/X_0'$, $y' := X_2'/X_0'$ and $z := X_0/X_0'$ such that

$$\begin{aligned} z &= X_0/X_0' = a_{00} + a_{01}x' + a_{02}y', \\ x &= X_1/X_0 = \frac{1}{z}(a_{10} + a_{11}x' + a_{12}y'), \\ y &= X_2/X_0 = \frac{1}{z}(a_{20} + a_{21}x' + a_{22}y'). \end{aligned}$$

Let f' be the defining polynomial of V in the X' variables, so that $f'(X') = f(X)$, and let $e := \deg V = \deg f = \deg f'$. Then

$$z^e \hat{f} = z^e \frac{f(X)}{X_0^e} = \frac{f'(X')}{X_0'^e} = \hat{f}'.$$

The following result was pointed out by D. Zelinsky in a private communication:

Lemma 4.5.12. *With the above notation, we have*

$$\frac{dx}{\hat{f}_y} = \det(A) z^{e-3} \frac{dx'}{\hat{f}'_{y'}}.$$

Proof. Suppose that $X' = A'X''$, where A' is another change of variable. Since $\det(AA') = \det(A)\det(A')$ and $X_0/X_0'' = (X_0/X_0')(X_0'/X_0'')$, the formula will hold for the change of variable AA' provided that it holds for A and A' separately. It therefore suffices to consider the following three cases:

Case 1: $x = ax' + by' + c$ $(a \neq 0), y = y', z = 1$. From the basic equation $\hat{f}'(x',y') = \hat{f}(ax' + by' + c, y')$, we obtain using the chain rule

$$\begin{aligned} \hat{f}'_{x'} &= a\hat{f}_x, \\ \hat{f}'_{y'} &= b\hat{f}_x + \hat{f}_y. \end{aligned}$$

Since $\hat{f}'(x',y')$ vanishes on V, we get

$$\hat{f}'_{x'} dx' = -\hat{f}'_{y'} dy'.$$

Using the above, we compute

$$\begin{aligned} dx = adx' + bdy' &= \frac{dx'}{\hat{f}_{y'}}(a\hat{f}'_{y'} - b\hat{f}'_{x'}) \\ &= a\frac{dx'}{\hat{f}_{y'}}(\hat{f}'_{y'} - b\hat{f}_x) \\ &= a\frac{dx'}{\hat{f}_{y'}}\hat{f}_y, \end{aligned}$$

as required.

Case 2: $x = y', y = x', z = 1$. This is immediate from the identity $dx/\hat{f}_y = -dy/\hat{f}_x$, since $\det A = -1$ in this case.

Case 3: $x = 1/x', y = y'/x', z = x'$. Note that $\det A = -1$ in this case as well. We have $\hat{f}'(x',y') = x'^e \hat{f}(1/x', y'/x')$ and thus $\hat{f}'_{y'} = x'^{e-1}\hat{f}_y$. Since $dx' = -x'^2 dx$, the result follows. □

In the language of projective maps, making a linear change of variables in $\mathbb{P}^n$ amounts to replacing one effective map with an equivalent effective map. Moreover, if $\phi = (1,x,y)$ and $\phi' = (1,x',y')$ with notation as in (4.5.12), then $[\phi] = [\phi'] + [z]$ because the function z defines the line at infinity in the X coordinates. We therefore have

Corollary 4.5.13. *Let $\phi = (1,x,y)$ be an effective plane map of degree e and let $\hat{f}(x,y)$ be the minimum polynomial satisfied by x and y. Then the divisor*

$$\Delta(\phi) := (e-3)[\phi] - \left[\frac{dx}{\hat{f}_y}\right]$$

depends only on the equivalence class of ϕ. □

We call Δ the *adjoint divisor* of the map ϕ and we put $\delta(\phi) := \deg\Delta(\phi)$. For a plane curve V with natural map ϕ, we write $\Delta(V) := \Delta(\phi)$ and $\delta(V) := \deg\Delta(V)$. We want to show that Δ is nonnegative. To see this, it is convenient to make a good choice of coordinates.

We say that a line L is *generic* with respect to the plane curve V if $\mu_a(V \cap L) = 1$ for all $a \in V \cap L$. This means that L is not tangent to V at any point and does not meet V at any singular points.

Lemma 4.5.14. *Let V be a plane curve. Then there exists a linear change of coordinates in $\mathbb{P}^2$ such that if $\phi = (1,x,y)$ is the natural map in this coordinate system then:*

1. *The lines $\mathbf{V}(X_i)$ $(i = 0,1,2)$ are generic.*
2. *The points $(1:0:0)$, $(0:1:0)$, and $(0:0:1)$ are not on V.*
3. *Both x and y are separating variables.*

Proof. Choose a point $a_0 \notin V$. The set of all lines through a_0 (usually called the *pencil* at a_0) is the set of all points on a line in the dual plane. This line meets the dual curve at finitely many points, which means that there are only finitely many tangents to V through a_0. Since there are only finitely many singular points, almost all lines through a_0 are generic with respect to V. Choose two distinct generic lines L_1 and L_2 through a_0, and choose a point $a_1 \neq a_0$ in $L_2 \setminus V$. Then there is a generic line $L_0 \neq L_2$ through a_1 that also misses every point of $L_1 \cap V$. Put $a_2 := L_0 \cap L_1$, and choose coordinates X_i so that $L_i = V(X_i)$ $(i = 0, 1, 2)$.

We now deal with the separability issue. Assume that $\mathrm{char}(k) = p > 0$, and recall that $u \in K$ is a separating variable if and only if $u \notin K^p$ (see (2.4.6)). We have $1 \in K^p$, and since $K = k(x,y)$, we cannot have $\langle \phi \rangle \subseteq K^p$, but it may happen that $\dim(\langle \phi \rangle \cap K^p) = 2$. This means that there is a point b on L_0 such that a linear form ℓ vanishes at b if and only if $\ell^* \in K^p$. To ensure that $b \neq a_1$, we can if necessary replace L_2 by another generic line on a_0 such that $L_2 \cap L_0 \notin V$. A similar adjustment is obviously possible in case $b = a_2$. □

We will refer to coordinates satisfying the above conditions as *generic coordinates*, and a projective map $(1,x,y)$ with respect to generic coordinates as a generic map. Note in particular that if $(1,x,y)$ is generic, the defining homogeneous polynomial f contains the monomials X_i^d with nonzero coefficients, which means that $\hat{f}$ is monic in both x and y.

We next show that $\Delta(\phi)$ is nonnegative and its support is precisely the set of singularities of ϕ. This result is basic to the theory of plane curves.

Theorem 4.5.15. *Let $\phi = (1,x,y)$ be a birational plane map and let $\hat{f}(x,y)$ be the minimum polynomial satisfied by x and y. Then $v_Q(\Delta(\phi)) \geq 0$ for all $Q \in \mathbb{P}_K$ with equality if and only if ϕ is nonsingular at Q.*

Proof. Fix a point $Q \in \mathbb{P}_K$. By virtue of (4.5.12), we may assume that ϕ is generic and finite at Q. As noted above, this implies that $\hat{f}$ is monic in x and y. In particular, y is integral over $k[x]$ and $\hat{f}$ is its minimum polynomial over $k(x)$. This means that $\hat{f}_y = \delta_{k(x)}(y)$. Since $K = k(x,y)$ by hypothesis, the desired inequality follows from (3.3.12).

Furthermore, if equality holds, we obtain $\mathscr{O}_P[y]_Q = \mathscr{O}_Q$. However, if we put $a := \phi(Q)$, then $k[x,y] \subseteq \mathscr{O}_a$ and $P = P_a \cap k[x]$, where P_a is the unique maximal ideal of $\mathscr{O}_a$. It is therefore clear that $\mathscr{O}_P[y] \subseteq \mathscr{O}_a$, and since $P_a = Q \cap \mathscr{O}_a$, we have $\mathscr{O}_P[y]_Q \subseteq \mathscr{O}_a$ as well. We conclude that $\mathscr{O}_a = \mathscr{O}_Q$, and thus ϕ is nonsingular at Q.

To complete the proof, assume that ϕ is nonsingular at Q, put $a := \phi(Q)$, and let $L_Q = \mathbf{V}(\ell_Q)$ be the unique line on a and $a_2 := (0:0:1)$. Since ϕ is finite at Q, we have $a = (1:\alpha:\beta)$ for some $\alpha, \beta \in k$. Then $L_Q = (-\alpha:1:0)$ and $\ell_Q^* := \ell_Q/X_0 = x - \alpha$. Recalling notation from (4.5.8), we see from (4.5.9) that $\hat{f}_x = f_1(1,x,y)$ and $\hat{f}_y = f_2(1,x,y)$. By (4.5.6), $v_Q(d\ell_Q^*) > 0$ if and only if L_Q is the tangent line at a, which occurs if and only if the the point a_2 is on the tangent line at a. Since $d\ell_Q^* = dx$, (4.5.7) yields $v_Q(dx) > 0$ if and only if $\hat{f}_y(Q) = 0$. Thus, if $v_Q(dx) = 0$ we have the required equality.

What happens if $\nu_Q(dx) > 0$? By the same argument with $a_1 := (0:1:0)$ replacing a_2, we obtain $\nu_Q(dy) > 0$ if and only if $\hat{f}_x(Q) = 0$. However, by (4.3.5) both conditions cannot hold simultaneously, because ϕ is nonsingular at Q. We conclude that if $\nu_Q(dx) > 0$, then $\nu_Q(dy) = 0 = \nu_Q(\hat{f}_x)$.

From (4.5.10) we get

$$\nu_Q(dx) + \nu_Q(\hat{f}_x) = \nu_Q(dy) + \nu_Q(\hat{f}_y).$$

Putting these together yields $\nu_Q(dx) = \nu_Q(\hat{f}_y)$, as required. □

Corollary 4.5.16. *Let V be a plane curve of degree d and genus g. Then*

$$g = \binom{d-1}{2} - \frac{1}{2}\delta(V).$$

In particular, $g \leq (d-1)(d-2)/2$ with equality if and only if V is nonsingular.

Proof. The formula is immediate from the definition of Δ, because $[dx/\hat{f}_y]$ is canonical of degree $2g-2$. □

It is immediate from the above result and (for example) (3.6.3) that there exist function fields with no nonsingular plane model. Indeed, we have

Corollary 4.5.17. *Let K be a function field of genus g. If K has a nonsingular plane model, then g is a triangular number.* □

A function $u \in K$ satisfies the *adjoint conditions* with respect to a plane map ϕ if $[u]_0 \geq \Delta(\phi)$. Such a function will be called an *adjoint function*. We proceed next to a study of the adjoint functions. The first step is

Lemma 4.5.18. *Let V be a plane curve with natural map $(1,x,y)$. Assume that y is integral over $k[x]$. Then*

$$k[x,y] = \cap\{\mathcal{O}_P[y] \mid P \in \mathbb{P}_{k(x)} \text{ and } \nu_P(x) \geq 0\}.$$

Proof. Obviously, $k[x] \subseteq \mathcal{O}_P$ and thus $k[x,y] \subseteq \mathcal{O}_P[y]$ for every P containing x. Conversely, let $K := k(x,y)$. Then every element $u \in K$ is uniquely a $k(x)$-linear combination

$$(*) \qquad u = \sum_{i=0}^{n-1} u_i y^i.$$

Because y is integral over $k[x]$, $\{1, y, \ldots, y^{n-1}\}$ is an $\mathcal{O}_P$-basis for $\mathcal{O}_P[y]$ for every prime P with $k[x] \subseteq \mathcal{O}_P$. Thus, $u \in \mathcal{O}_P[y]$ implies $u_i \in \mathcal{O}_P$ for all i by uniqueness of $(*)$. Since $k[x] = \cap\{\mathcal{O}_P \mid x \in P\}$, the result follows. □

For the remainder of this section we let V be a plane curve with natural map $\phi = (1,x,y)$, function field $K = k(x,y)$, minimum polynomial $\hat{f}(x,y)$, and we put $\omega(V) := [dx/\hat{f}_y]$. From (3.3.10) and (3.3.11), we see that the local conductor $C_P(y)$ is given by

$$C_P(y) = \{u \in K \mid \nu_{Q_i}(u) \geq \nu_{Q_i}(\Delta(V))\ (1 \leq i \leq r)\},$$

where $P \in \mathbb{P}_{k(x)}$ and $Q_1, \ldots, Q_r$ are the prime divisors of P in K. We therefore define the *conductor* of V as

$$C(V) := \{u \in K \mid v_Q(u) \geq v_Q(\Delta(V)) \text{ for all finite } Q\}.$$

This terminology is justified by Exercise 4.1.

Let $R(V)$ denote the integral closure of $k[V] = k[x,y]$ in K. Then

$$R(V) = \cap\{\mathscr{O}_Q \mid Q \text{ finite}\}$$

by (1.1.8). It follows that $C(V)$ is an ideal of $R(V)$ consisting of all adjoint functions in $R(V)$, or what is the same thing, the set of all adjoint functions with no poles in the finite plane. We see that all such functions lie in $k[x,y]$. In particular, it follows that $L([\omega(V)]) \subseteq k[x,y]$. By abuse of terminology, we will refer to elements of $k[x,y]$ as "polynomials." By the *degree* of an element $h \in k[x,y]$ we mean the minimal degree of a polynomial $h(X,Y) \in k[X,Y]$ with $h(x,y) = h$. For any nonnegative integer i, let $k[V]_i$ denote the set of all polynomials of degree at most i.

Note that $L([\omega(V)])$ is the set of all adjoint functions all of whose poles lie at infinity and have order at most $d-3$. This certainly includes all adjoint polynomials of degree at most $d-3$. Indeed, when V is nonsingular, we have $\delta(V) = 0$, and a dimension count shows that

$$\dim k[V]_{d-3} = \binom{d-1}{2} = g = \dim L([\omega(V)]).$$

What happens in the singular case? The key fact that we need here is that a polynomial all of whose poles have order at most $d-3$ in fact has degree at most $d-3$. This follows from

Theorem 4.5.19. *Let $V = V(f)$ be a plane curve of degree d in generic coordinates with natural map $(1,x,y)$, and let $h \in k[x,y]$. If all poles of h have order at most e for some $e \in \mathbb{Z}$, then h has degree at most e.*

Proof. Put $K := k(V)$ and let $\{a_1, \ldots, a_d\} \in V$ be the d distinct points at infinity. Since $(0:0:1) \notin V$, we have $a_i = (0:1:\alpha_i)$ for distinct $\alpha_1, \alpha_2, \ldots, \alpha_d \in k$. Let $Q_i \in \mathbb{P}_K$ with $\phi(Q_i) = a_i$. Then $1/x$ is a local parameter at Q_i for all i, since x and y have simple poles at Q_i. It follows from the definition of $\phi(Q)$ that if we put $y_1 := y/x$, then $\alpha_i = y_1(Q_i)$. This implies that the Laurent expansion of y at Q_i is

$$y = \alpha_i x + \alpha_{i0} + \alpha_{i1}(1/x) + \cdots,$$

where the higher-order coefficients α_{ij} are irrelevant. In particular, we conclude that $y - \alpha_i x$ is the unique linear function of x and y that has no pole at Q_i.

For $h(X,Y) \in k[X,Y]$, we put $h^* := h(x,y) \in K$. If $\deg h = e$, we can write

$$h(X,Y) = \sum_{j=0}^{e} h_j(X,Y),$$

where h_j is homogeneous of degree j and $h_e \neq 0$. Fix an index i. If $v_{Q_i}(h_e^*) = -e$, then $v_{Q_i}(h^*) = -e$ because $v_{Q_i}(h_j^*) \geq -j$ for all j. Thus, if $v_{Q_i}(h^*) > -e$, we conclude that h_e is divisible by $Y - \alpha_i X$.

It follows that all poles of h are of order less than e if and only if h_e is divisible by the d distinct linear forms $Y - \alpha_i X$, $1 \leq i \leq d$. This holds in particular for $h = \hat{f}(X,Y)$ and $e = d$, because $\hat{f}^* = 0$, whence $\hat{f}_d^*$ is a sum of polynomials of degree less than d. We conclude that

$$\hat{f}_d(X,Y) := \alpha \prod_{i=1}^{d} (Y - \alpha_i X)$$

for some scalar α.

If now, by way of contradiction, h has degree e but all poles of h^* are of order less than e, we get $h_e = u(X,Y)\hat{f}_d$ for some homogeneous u of degree $e-d$, but then

$$\tilde{h}(X,Y) := h(X,Y) - u(X,Y)\hat{f}(X,Y)$$

has degree less than e and $h^* = \tilde{h}^*$ because $\hat{f}^* = 0$. This violates our definition of degree and completes the proof. □

Corollary 4.5.20. *With the above notation,* $C(V) \subseteq k[V]$, *and*

$$L([\omega(V)]) = C(V) \cap k[V]_{d-3}.$$

In particular, a differential form ω *is regular if and only if it is of the form*

$$\omega = \frac{p(x,y)}{\hat{f}_y} dx,$$

where $p(x,y)$ *is an adjoint polynomial of degree at most* $d-3$.

Proof. From the definitions, we see that

$$C(V) = \cap\{C_P(y) \mid P \in \mathbb{P}_{k(x)} \text{and } v_P(x) \geq 0\}.$$

Thus, (4.5.18) implies that $C(V) \subseteq k[V]$.

By (4.5.16) and the definitions, $L([\omega(V)]) \subseteq C(V)$. Since $\omega(V)$ vanishes to order exactly $d-3$ at each point at infinity, every element $u \in L([\omega(V)])$ has poles of order at most $d-3$, and therefore degree at most $d-3$ by (4.5.19). □

From (4.5.20) we see that the binomial coefficient in the genus formula (4.5.16) is no accident. Namely, since there are no polynomial relations on x and y of degree less than d, $\dim k[V]_{d-3} = \binom{d-1}{2}$. On the other hand, since the dimension of $L([\omega(V)])$ is g, we see from the genus formula that $\delta(V)/2$ is the dimension of the space of linear constraints imposed by the adjoint conditions on the space of polynomials of degree $d-3$. This is a version of the so-called *Gorenstein relations* [9]. However, there is more to be said.

Theorem 4.5.21. *Let V be a plane curve with generic coordinates. Then*

$$\dim_k(R(V)/k[V]) = \dim_k(k[V]/C(V)) = \frac{1}{2}\delta(V).$$

Moreover, $k[V] = k[V]_{d-3} + C(V)$.

Proof. Put $L := L((d-3)[\phi])$. This is the space of functions that are regular in the finite plane and have poles of order at most $d-3$. Thus, $L \subseteq R(V)$, and since we have generic coordinates, $L \cap k[V] = k[V]_{d-3}$ by (4.5.19).

Note that $\deg((d-3)[\phi]) = d(d-3) = 2g-2+\delta(V)$. If $\delta(V) = 0$, we have $C(V) = k[V] = R(V)$, so we may as well assume that $\delta(V) > 0$, and thus $(d-3)[\phi]$ is nonspecial.

By (4.5.20), we have $L([\omega(V)]) = L \cap C(V)$, and therefore, as discussed above, (4.5.16) yields

$$(*)\quad \begin{aligned} \frac{\delta(V)}{2} &= \dim_k k[V]_{d-3}/(k[V]_{d-3} \cap C(V)) = \dim_k (k[V]_{d-3} + C(V))/C(V) \\ &\leq \dim_k k[V]/C(V), \end{aligned}$$

with equality if and only if $k[V] = k[V]_{d-3} + C(V)$.

From the Riemann–Roch theorem and (4.5.16) we have

$$\begin{aligned} \dim_k L = d(d-3) - g + 1 &= \binom{d-1}{2} + \frac{\delta(V)}{2} - 1 \\ &= \dim_k(k[V]_{d-3}) + \frac{\delta(V)}{2} - 1, \end{aligned}$$

because $(d-3)[\phi]$ is nonspecial. Since $L \cap k[V] = k[V]_{d-3}$, we get

$$(**)\quad \begin{aligned} \frac{\delta(V)}{2} - 1 &= \dim_k(L/(L \cap k[V])) = \dim_k(L + k[V])/k[V] \\ &\leq \dim_k R(V)/k[V], \end{aligned}$$

with equality if and only if $R(V) = L + k[V]$. In fact, we claim that $L + k[V]$ is a proper subspace of $R(V)$. Namely, consider the k-linear functional

$$\mu(h) := \sum_{i=1}^{r} \operatorname{Res}_{Q_i}(h\omega),$$

where $\omega := \omega(V)$ has poles $\{Q_1, \ldots, Q_r\}$. For $h \in L$, $h\omega$ is regular at infinity by (4.5.16) and the definition of L, and has finite poles only at the poles of ω. Thus, $\mu|_L = 0$ by the residue theorem (2.5.4).

For $h \in k[V]$ we have

$$\operatorname{tr}_{K/k(x)} \frac{h}{\hat{f}_y} \in k[x]$$

by (3.3.9), where $\hat{f}(x,y)$ is the dehomogenized defining polynomial of V as before. Therefore, another application of the residue theorem gives

$$\mathrm{Res}_\infty \mathrm{tr}_{K/k(x)}\left(\frac{h}{\hat{f}_y}\right) dx = 0,$$

where ∞ is the unique infinite prime of $k(x)$. Now the trace formula (3.1.5) together with a final application of the residue theorem yields

$$\mu(h) = -\sum_{Q|\infty} \mathrm{Res}_Q(h\omega) = \mathrm{Res}_\infty \mathrm{tr}_{K/k(x)}\left(\frac{h}{\hat{f}_y}\right) dx = 0.$$

We have shown that μ vanishes on $L + k[V]$, but by the strong approximation theorem (2.2.13) there is an element $h \in R(V)$ such that $h\omega$ has a simple pole at Q_1 and is regular at Q_i for $i > 1$, whence $\mu(h) \neq 0$. We conclude that $L + k[V] \subsetneq R(V)$, and therefore $(**)$ yields

$$\dim_k(R(V)/k[V]) \geq \frac{\delta(V)}{2}.$$

Comparing with $(*)$, we see that to complete the proof, it will suffice to show that

$$(***) \qquad \dim_k(R(V)/C(V)) \leq \delta(V).$$

For any divisor D, define

$$I(D) := \{u \in R(V) \mid [u]_0 \geq D\}.$$

If t is a local parameter at Q and $v_Q(D) = e$, the map $\eta(u) = t^{-e}u + Q$ defines a k-linear map $\eta : I(D) \to \mathscr{O}_Q/Q = k$ with $\ker \eta = I(D+Q)$. It follows that $\dim_k I(D+Q) \leq \dim_k I(D) + 1$ for any divisor D.

Now, if we choose a chain of divisors

$$0 = D_0 \leq D_1 \leq \cdots \leq D_{\delta(V)} = \Delta(V),$$

where $\deg D_i = i$, we get a corresponding chain of ideals

$$R(V) = I(0) \supseteq I(D_1) \supseteq \cdots \supseteq I(D_{\delta(V)}) = C(V),$$

and $(***)$ follows. □

4.6 Exercises

Exercise 4.1. Let $\phi = (1,x,y) : \mathbb{P}_K \to V$ be a generic projective map and let $P \in \mathbb{P}_{k(x)}$ be finite.

(i) Let R_P denote the integral closure of $\mathscr{O}_P$ in K, as in (3.3.10). Show that R_P is the localization of $R(V)$ at the set of prime divisors of P in K. [Hint: (2.2.13).]

(ii) Prove that $C(V)$ is the largest ideal of $R(V)$ contained in $k[V]$.

Exercise 4.2. Let k be algebraically closed with $\mathrm{char}(k) \neq 2$ and suppose that $K = k(x,y)$ where $y^2 = f(x)$ for some square-free polynomial $f(X)$ of degree d.

(i) If $d = 2m$ is even, show that after a change of variable

$$\tilde{x} = 1/(x-a), \quad \tilde{y} = y/(x-a)^m$$

for suitable a, we have $\tilde{y}^2 = \tilde{f}(\tilde{x})$, where $\tilde{f}$ is square-free of degree $d-1$. Changing notation, we may assume that d is odd. Replacing x by an additive translate if necessary, assume in addition that $f(0) \neq 0$.

(ii) Show that the map $\phi := (1,x,y)$ is nonsingular in the finite plane and singular at infinity for $d \geq 4$.

(iii) Show that the change of variable $\tilde{x} := x/y$, $\tilde{y} = 1/y$ yields a generic projective map $\tilde{\phi} = (1,\tilde{x},\tilde{y})$ with defining equation

$$g(\tilde{x},\tilde{y}) := \tilde{y}^{d-2} - f(\tilde{x}/\tilde{y})\tilde{y}^d,$$

and that $\tilde{\phi}$ is singular at exactly one point $Q \in \mathbb{P}_K$, namely where $\tilde{\phi}(Q) = (1:0:0)$.

(iv) Show that $v_Q(\tilde{x}) = d-2$, $v_Q(\tilde{y}) = d$, $v_Q(d\tilde{y}) = d-1$, and $v_Q(g_{\tilde{x}}) = (d-1)(d-2)$.

(v) Compute $\delta(V)$ and then g_K using (4.5.16). Compare with (3.6.3).

Exercise 4.3. Let k be algebraically closed of characteristic 2 and let $K = k(x,y)$ with $y^2 + y = x^{2g+1}$ for some positive integer g. Show that K has exactly one Weierstrass point of weight $(g-1)g(g+1)$.

Exercise 4.4. Let k_0 be a subfield of an algebraically closed field k. If $f \in k_0[X_0,X_1,X_2]$ is homogeneous and irreducible over k, we say that $V := V(f)$ is *defined over* k_0. Let $K := k(V)$ with natural map $\phi := (1,x,y)$. Let $K_0 := k_0(x,y)$. We say that a point $(a:b:c) \in \mathbb{P}^2$ is defined over k_0 if $\{\lambda a, \lambda b, \lambda c\} \subseteq k_0$ for some nonzero $\lambda \in k$. Show that if a point $P \in \mathbb{P}_K$ is defined over k_0 (see (3.2.7)) then $\phi(P)$ is defined over k_0. Conversely, if ϕ is nonsingular at P and $\phi(P)$ is defined over k_0, show that P is defined over k_0.

Exercise 4.5. Let q be a power of a prime p, let k be algebraically closed of characteristic p, and let $V = V(X_0^{q+1} + X_1^{q+1} + X_2^{q+1})$ with natural map ϕ. Prove the following:

(i) V is nonsingular. The dual curve $\tilde{V}$ has the same defining equation as V, but the map $\tilde{\phi}$ of (4.5.11) is not birational.

(ii) Identify the points of K with their images under ϕ. Then

$$[\phi] = \sum_{b^{q+1}=-1} (0:1:b).$$

(iii) As usual, put $x := X_1/X_0$ and $y := X_2/X_0$, then

$$[y]_0 = \sum_{a^{q+1}=-1} (1:a:0), \quad \text{and}$$
$$[dx] = q[y]_0 - 2[\phi].$$

(iv) The matrix of Hasse derivatives with respect to x is

$$\begin{bmatrix} 1 & 0 & 0 & \dots & 0 & \dots \\ x & 1 & 0 & \dots & 0 & \dots \\ y & -(x/y)^q & 0 & \dots & D_x^{(q)}(y) & \dots \end{bmatrix}.$$

[Hint: Apply $D_x^{(i)}$ to the defining equation and use Exercise 1.13.]

(v) The order sequence of ϕ is $0, 1, q$ and the Wronskian is

$$W_x(\phi) = D_x^{(q)}(y) = \frac{yx^{q^2} - xy^{q^2}}{y^{q^2+q}}.$$

(vi) There are exactly $q^3 - q$ points in the finite plane that are defined over $GF(q)$, and they are just the zeros of $w := yx^{q^2} - xy^{q^2}$.

(vii) Each point at infinity is defined over $GF(q^2)$ and is a pole of w of order $q^2 - q$. The zeros of w are all simple. [Hint: Let $u := 1/x$ and $v := y/x$.]

(viii) For any point P we have $q \le j_2(P) \le q+1$. Every Weierstrass point of ϕ has weight 1. [Hint: (4.4.21).]

(ix) The Weierstrass divisor is $R_\phi = [w]_0 + [\phi]$, the sum of all $GF(q^2)$-rational points of V.

(x) For $q = 3$, ϕ is the canonical map and V is nonclassical. The Weierstrass gap sequence is $1, 2, 4$.

Exercise 4.6. Let $\phi := (\phi_0, \dots, \phi_n)$ be an effective projective map that is normalized at P and let $J(\phi)(P) = (j'_1, \dots, j'_n)$. Choose a local parameter t at P and let $h_{ij}(P)$ be the scalars defined in (4.4.6). Show that the equation of the osculating hyperplane at P is

$$\det \begin{bmatrix} X_0 & h_{00}(P) & h_{0j'_1}(P) & \dots & h_{0j'_{n-1}}(P) \\ X_1 & h_{10}(P) & h_{1j'_1}(P) & \dots & h_{1j'_{n-1}}(P) \\ \vdots & \vdots & \vdots & \dots & \vdots \\ X_n & h_{n0}(P) & h_{nj'_1}(P) & \dots & h_{nj'_{n-1}}(P) \end{bmatrix} = 0.$$

[Hint: Use (4.4.13).]

5
Zeta Functions

In this chapter the ground field k will be finite of characteristic p and order $q := p^r$, and therefore, of course, perfect. We are interested in counting the number of points of a function field K/k and all its scalar extensions, but for reasons that will be evident shortly, we consider instead the related quantity

$$a_K(n) := |\{D \in \mathrm{Div}(K) \mid D \geq 0 \text{ and } \deg D = n\}|,$$

which, as we will show, is finite; and we define

$$Z_K(t) := \sum_{n=0}^{\infty} a_K(n) t^n.$$

Note that $a_K(1)$ is the number of points of K. As we will see below, $Z_K(t)$ has radius of convergence $1/q$ in the complex plane and so defines an analytic function there, called the *zeta function* of K, for the following reason. Given any nonnegative divisor D, define the *absolute norm* $N(D) := q^{\deg D}$, and put

$$\zeta_K(s) := \sum_{D \geq 0} N(D)^{-s}.$$

The function ζ_K bears an obvious resemblance to the classical Riemann zeta function, and we have

$$\zeta_K(s) = \sum_{n=0}^{\infty} a_K(n) q^{-ns} = Z_K(q^{-s}).$$

In fact, the resemblance is more than superficial. We will show that $\zeta_K(s)$ has an Euler product representation

$$\zeta_K(s) = \prod_{P \in \mathbb{P}_K} (1 - N(P)^{-s})^{-1}$$

and satisfies the functional equation

$$\zeta_K(1-s) = N(C)^{s-1/2}\zeta_K(s),$$

where $N(C) = q^{2g_K-2}$ is the absolute norm of the canonical class. Unlike the Riemann zeta function, however, $Z_K(t)$ turns out to be a rational function

$$Z_K(t) = \frac{L_K(t)}{(1-t)(1-qt)},$$

where $L_K(t)$ is a polynomial of degree $2g_K$. The main goal of this chapter is to prove that the roots of $L(t)$ are of absolute value $q^{-1/2}$, which is equivalent to the statement that the zeroes of $\zeta_K(s)$ lie on the line $\Re(s) = \frac{1}{2}$. This is of course the analogue of the classical Riemann Hypothesis, and was first proved by Weil [23]. Our proof is due to Stöhr–Voloch [19] and Bombieri [2]. As a consequence, we get a very powerful estimate for the number of points of K.

5.1 The Euler Product

We denote by k_r the unique extension of k of degree r (see (A.0.19)), and for a function field K/k we write K_r for the unique scalar extension $k_r \otimes_k K$ of K of degree r (see Section 3.2).

Let $P \in \mathbb{P}_K$. If we choose a subfield $k(x) \subseteq K$ with $|K : k(x)| = m$, then $\deg P \leq m \deg(P \cap k(x))$, so a prime divisor of K of degree at most n divides an irreducible polynomial of $k(x)$ of degree at most mn. Moreover, at most m distinct primes P can divide the same irreducible polynomial. Since there are only finitely many irreducible polynomials of degree at most mn, it follows that there are only finitely many prime divisors of K of degree at most n. In particular, the infinite product

$$Z_K(t) := \prod_{P \in \mathbb{P}_K} (1 - t^{\deg P})^{-1} \tag{5.1.1}$$

makes sense as a formal product of formal power series, and the coefficient of t^n is

$$a_K(n) := |\{D \in \mathrm{Div}(K) \mid D \geq 0 \text{ and } \deg D = n\}|.$$

Moreover, if θ is an rth root of unity in the complex plane, the infinite product

$$\prod_{P \in \mathbb{P}_K} (1 - (\theta t)^{\deg P})^{-1}$$

also makes sense as a formal product. Here the coefficient of t^n is a finite integral combination of rth roots of unity. We will shortly prove that (5.1.1) in fact has positive radius of convergence $r = q^{-1}$ in the complex plane. First, however, we obtain an important relationship between $Z_K(t)$ and $Z_{K_n}(t)$, where K_n is the unique scalar extension of K of degree n. We need the following polynomial identity.

Lemma 5.1.2. *Let d and n be positive integers. Then*

$$\left(1 - t^{nd/(n,d)}\right)^{(n,d)} = \prod_{\theta^n=1} 1 - (\theta t)^d,$$

where $(n,d) := \gcd(n,d)$ and the product is over all complex n^{th} roots of unity.

Proof. The basic identity is

$$1 - t^n = \prod_{\theta^n=1} 1 - \theta t. \tag{5.1.3}$$

If θ is a primitive n^{th} root of unity, then $\mu := \theta^d$ is a primitive $n/(n,d)^{\text{th}}$ root of unity, and we have

$$\prod_{\theta^n=1} 1 - (\theta t)^d = \left(\prod_{\mu^{n/(n,d)}=1} 1 - \mu t^d\right)^{(n,d)} = (1 - t^{nd/(n,d)})^{(n,d)}. \quad \square$$

Given a prime P of K and a divisor Q of P in K_n, we have $e(Q|P) = 1$ by (3.2.3). Let $d := \deg P$. Then F_P is the unique extension of k of degree d, and by (3.2.3) it follows that $F_Q = F_P k_n$. Since $k_n \cap F_P$ is the unique extension of k of degree (n,d), we get $\deg Q = |F_Q : k_n| = |F_P : k_{(n,d)}| = d/(n,d)$ and $f(Q|P) = n/(n,d)$. Furthermore, if the number of primes Q dividing P is r, then $n = rn/(n,d)$ by (2.1.17), so $r = (n,d)$. Now using the lemma we get

$$\prod_{Q:Q|P} (1 - t^{n \deg Q}) = (1 - t^{nd/(n,d)})^{(n,d)} = \prod_{\theta^n=1} 1 - (\theta t)^d.$$

This implies that

$$\begin{aligned} Z_{K_n}(t^n) &= \prod_{P \in \mathbb{P}_K} \prod_{Q:Q|P} (1 - t^{n \deg Q})^{-1} = \prod_{P \in \mathbb{P}_K} \prod_{\theta^n=1} (1 - (\theta t)^{\deg P})^{-1} \\ &= \prod_{\theta^n=1} Z_K(\theta t). \end{aligned} \tag{5.1.4}$$

Equation (5.1.4) is quite powerful, but to use it, we first need to show that $Z_K(t)$ is a rational function. The Riemann–Roch theorem says that for large n, every divisor class of degree n contains a nonnegative divisor. Since $a_K(n)$ is finite, it follows that there are only finitely many divisor classes of degree n for large n. But the divisor classes of degree n form a coset of the degree-zero subgroup, $J(K)$, of the divisor class group. We conclude that $h_K := |J(K)| < \infty$. The integer h_K is called the *class number* of K. We will drop the subscript when there is no danger of confusion.

The next issue that arises is that the degree map may not be surjective. We will prove shortly that it is, but for the time being we let r denote the index of the degree map in $\mathbb{Z}$. So the number of divisor classes of degree n is zero unless $r \mid n$, in which case it is equal to the class number h.

Finally, we count the number of nonnegative divisors in a divisor class. Since $x \in L(D)$ iff $D + [x] \geq 0$ and since $[x] = [\alpha x]$ for any nonzero scalar α, the number of nonnegative divisors linearly equivalent to D is just the number of 1-dimensional subspaces of $L(D)$. Namely, we have

$$|\{D' \in \bar{D} \mid D' \geq 0\}| = \frac{q^{l(d)} - 1}{q - 1}, \tag{5.1.5}$$

where $\bar{D}$ denotes the divisor class of D and $l(D) := \dim L(D)$. Since we have $\dim L(D) = n - g + 1$ for $n \geq 2g - 1$ by Riemann–Roch, we have proved

Lemma 5.1.6. *Let K be a function field over a finite field k of order q. Then the Jacobian of K has finite order h, and if r denotes the index of the image of the degree map* $\mathrm{Div}(K) \to \mathbb{Z}$*, then for $n \geq 2g - 1$ we have*

$$a(n) = \begin{cases} h\frac{q^{n-g+1}-1}{q-1} & \text{if } r \mid n, \\ 0 & \text{otherwise.} \end{cases} \qquad \square$$

It follows that $Z_K(t)$ is a rational function. Namely,

$$\begin{aligned} Z_K(t) &= \sum_{nr \leq 2g-2} a_K(nr)t^{nr} + \sum_{nr \geq 2g-1} h\frac{q^{nr-g+1} - 1}{q - 1}t^{nr} \\ &= F(t) + h\frac{q^{1-g}}{q - 1} \sum_{nr \geq 2g-1} (qt)^{nr} - h\frac{1}{q - 1} \sum_{nr \geq 2g-1} t^{nr} \\ &= \frac{L(t)}{(1 - (qt)^r)(1 - t^r)}, \end{aligned} \tag{5.1.7}$$

where $F(t)$ and $L(t)$ are polynomials.

Theorem 5.1.8. *Let K be a function field over a finite field k of order q. Then the degree map* $\deg : \mathrm{Div}(K) \to \mathbb{Z}$ *is onto.*[1] *Thus,*

$$Z_K(t) = \frac{L_K(t)}{(1 - t)(1 - qt)} \tag{5.1.9}$$

for some polynomial $L_K(t)$.

Proof. Let r be the index of the image of the degree map as in (5.1.7) above. We have $Z_K(\theta t) = Z_K(t)$ for θ an rth root of unity, so (5.1.4) becomes

$$Z_{K_r}(t^r) = Z_K(t)^r. \tag{$*$}$$

There are infinitely many prime divisors of K, because at least one prime of K divides each prime of $k(x)$ and there are infinitely many prime divisors of $k(x)$. This implies that infinitely many coefficients in the power series expansion of

[1] In general, the degree map is not onto. See Exercise 5.3.

$Z_K(t)$ around $t = 0$ are positive. Since all coefficients are nonnegative, $Z_K(t)$ diverges at $t = 1$, and therefore $Z_K(t)$ has at least one pole in the unit disk. By (*) $Z_{K_r}(t^r)$ has a pole of order at least r. On the other hand, (5.1.7) shows that all poles of $Z_K(t)$ are simple, and therefore the poles of $Z_{K_r}(t^r)$ are simple as well. We conclude that $r = 1$. □

5.2 The Functional Equation

Now that we know the degree map is onto, the formula for $a(n)$ is simplified, and we can prove

Theorem 5.2.1. *Let K be a function field over a finite field k of order q. Then $L_K(1) = h_K$. Moreover, the zeta function satisfies the functional equation*

$$Z_K(t) = q^{g-1}t^{2g-2}Z_K\left(\frac{1}{qt}\right).$$

Proof. For $g = 0$ we have $h = 1$ and $l(D) = \deg D + 1$ for all $D \geq 0$, whence

$$\begin{aligned} Z_K(t) &= \sum_{n=0}^{\infty} \frac{q^{n+1}-1}{q-1} t^n = \frac{1}{q-1}\left(\frac{q}{1-qt} - \frac{1}{1-t}\right) \\ &= \frac{1}{(1-qt)(1-t)}, \end{aligned}$$

and the theorem is easily verified in this case. For $g > 0$ we let $\bar{D}$ denote the divisor class of D, and we let $\mathscr{S} := \{\bar{D} \mid 0 \leq \deg D \leq 2g-2\}$. Then

$$\begin{aligned} Z_K(t) &= \sum_{\bar{D}\in\mathscr{S}} \frac{q^{l(D)}-1}{q-1} t^{\deg\bar{D}} + \sum_{n=2g-1}^{\infty} h\frac{q^{n-g+1}-1}{q-1} t^n \\ &= \frac{1}{q-1}\sum_{\bar{D}\in\mathscr{S}} q^{l(D)}t^{\deg\bar{D}} + \frac{hq^{1-g}}{q-1}\sum_{n=2g-1}^{\infty} q^n t^n - \frac{h}{q-1}\sum_{n=0}^{\infty} t^n \qquad (5.2.2) \\ &= \frac{1}{q-1}\sum_{\bar{D}\in\mathscr{S}} q^{l(D)}t^{\deg\bar{D}} + \frac{h}{q-1}\left(\frac{q^g t^{2g-1}}{1-qt} - \frac{1}{1-t}\right). \end{aligned}$$

Using $L_K(t) = (1-t)(1-qt)Z_K(t)$, (5.2.2) yields $L_K(1) = h$.

We verify the functional equatiown for the sum over the low-degree divisors using the full strength of the Riemann–Roch theorem:

$$\begin{aligned}
\sum_{\bar{D}\in\mathcal{S}} q^{l(D)}(qt)^{-\deg D} &= \sum_{\bar{D}\in\mathcal{S}} q^{\deg D-g+1+l(C-D)}(qt)^{-\deg D}\\
&= (qt)^{2-2g}\sum_{\overline{C-D}\in\mathcal{S}} q^{\deg D-g+1+l(C-D)}(qt)^{\deg(C-D)}\\
&= (qt)^{2-2g}\sum_{\overline{C-D}\in\mathcal{S}} q^{\deg D+\deg(C-D)-g+1+l(C-D)}t^{\deg(C-D)}\\
&= (qt)^{2-2g}\sum_{\bar{D}\in\mathcal{S}} q^{g-1+l(D)}t^{\deg D}\\
&= q^{1-g}t^{2-2g}\sum_{\bar{D}\in\mathcal{S}} q^{l(D)}t^{\deg D},
\end{aligned}$$

where C is a canonical divisor, and we have used the fact that as D ranges over all nonnegative divisors of degree at most $2g-2$, so does $C-D$. We therefore have

$$\begin{aligned}
Z_K\left(\frac{1}{qt}\right) &= \frac{1}{q-1}\sum_{\bar{D}\in\mathcal{S}} q^{l(D)}(qt)^{-\deg D} + \frac{h}{q-1}\left(\frac{q^g(qt)^{1-2g}}{1-\frac{q}{qt}} - \frac{1}{1-\frac{1}{qt}}\right)\\
&= \frac{q^{1-g}t^{2-2g}}{q-1}\sum_{\bar{D}\in\mathcal{S}} q^{l(D)}t^{\deg D} + \frac{h}{q-1}\left(\frac{qt}{1-qt} - \frac{q^{1-g}t^{2-2g}}{1-t}\right)\\
&= q^{1-g}t^{2-2g}\left(\frac{1}{q-1}\sum_{\bar{D}\in\mathcal{S}} q^{l(D)}t^{\deg D} + \frac{h}{q-1}\left(\frac{q^g t^{2g-1}}{1-qt} - \frac{1}{1-t}\right)\right)\\
&= q^{1-g}t^{2-2g}Z_K(t). \quad \square
\end{aligned}$$

Making the substitution $t = q^{-s}$, we immediately get

Corollary 5.2.3. *If C is the canonical class, then*

$$\zeta_K(s) = N(C)^{s-1/2}\zeta_K(1-s). \quad \square$$

For the numerator of the zeta function, we have

Corollary 5.2.4. *$L_K(t)$ satisfies the functional equation*

$$L_K(t) = q^g t^{2g} L_K\left(\frac{1}{qt}\right).$$

In particular, $\deg L_K(t) = 2g$ and if $L_K(t) = \sum_{i=0}^{2g} a_i t^i$, then

$$a_{2g-i} = q^{g-i}a_i \quad \textit{for all } i.$$

Proof. The functional equation for $L_K(t)$ follows easily from (5.1.9) and the functional equation for $Z_K(t)$. From it we get

$$\sum_{i=0}^{n} a_i t^i = q^g t^{2g}\sum_{i=0}^{n} a_i (qt)^{-i} = \sum_{i=0}^{n} a_i q^{g-i}t^{2g-i}.$$

This implies that $n = 2g$ and that $a_i q^{g-i} = a_{2g-i}$ for all i. □

Corollary 5.2.5. *There exist algebraic integers* $\{\alpha_1, \ldots, \alpha_{2g}\}$ *such that*

$$L_K(t) = \prod_{i=1}^{2g}(1 - \alpha_i t)$$

and $\alpha_i \alpha_{2g-i+1} = q$ *for* $1 \le i \le g$.

Proof. From (5.1.9) we have $L_K(t) = (1-t)(1-qt)Z_K(t)$. It follows that $L_K(t)$ has integer coefficients and constant term equal to 1. By (5.2.4) the leading coefficient of $L_K(t)$ is q^g. Thus, the reciprocal polynomial is monic with constant term q^g, so we can write $L_K(t) = \prod_{i=1}^{2g}(1 - \alpha_i t)$ where the α_i are algebraic integers and $\prod_{i=1}^{2g} \alpha_i = q^g$. Now the functional equation yields

$$\begin{aligned} L_K(t) = q^g t^{2g} L_K\left(\frac{1}{qt}\right) &= q^g t^{2g} \prod_{i=1}^{2g}\left(1 - \frac{\alpha_i}{qt}\right) \\ &= \frac{1}{q^g} \prod_{i=1}^{2g}(qt - \alpha_i) = \prod_{i=1}^{2g}\left(1 - \frac{q}{\alpha_i}t\right). \end{aligned}$$

This means that every α_j is equal to some q/α_i. The problem is that we might have $i = j$, that is, $\alpha_i^2 = q$. However, since $\deg L(t)$ is even, the total number of such α_i is even, and since the product of all the α_i is positive, the number of α_i equal to $-\sqrt{q}$ is even, and therefore so is the number equal to $+\sqrt{q}$. This implies that notation can be chosen so that $\alpha_i \alpha_{2g-i+1} = q$ for all i. □

Corollary 5.2.6. *Let* $L_K(t) = \prod_{i=1}^{2g}(1 - \alpha_i t)$. *Then* $L_{K_n}(t) = \prod_{i=1}^{2g}(1 - \alpha_i^n t)$.

Proof. This is straightforward using (5.1.4) and the identity (5.1.3). □

5.3 The Riemann Hypothesis

We now prove that the zeros of $Z_K(t)$ have absolute value $q^{-1/2}$. Note that $|q^{-s}| = q^{-\Re(s)}$, so we are proving that the zeros of $\zeta_K(s)$ all lie on the line $\Re(s) = \frac{1}{2}$.

Corollary 5.3.1. *The Riemann Hypothesis holds for K if and only if it holds for some scalar extension* K_n.

Proof. Since the zeros of $Z_{K_n}(t)$ are just the nth powers of the zeros of $Z_K(t)$ by (5.2.6), we have $|\alpha| = q^{1/2}$ if and only if $|\alpha^n| = q^{n/2}$. □

We next compute the logarithmic derivative of $Z_K(t)$:

$$\mathscr{L}_K(t) := \frac{Z_K'(t)}{Z_K(t)} = \sum_{i=1}^{2g} \frac{-\alpha_i}{1-\alpha_i t} + \frac{1}{1-t} + \frac{q}{1-qt} \tag{5.3.2}$$
$$= \sum_{n=0}^{\infty} \left[1 + q^{n+1} - \sum_{i=1}^{2g} \alpha_i^{n+1}\right] t^n.$$

Equating constant terms, we have

$$a_K(1) = 1 + q - \sum_{i=1}^{2g} \alpha_i. \tag{5.3.3}$$

More generally, define

$$b_K(n) := 1 + q^n - \sum_{i=1}^{2g} \alpha_i^n.$$

From (5.2.6) we have $b_K(n) = a_{K_n}(1)$. Summarizing all of this, we have proved

Theorem 5.3.4. *Let K be a function field over a finite field k of order q, and let*

$$L_K(t) = \prod_{i=1}^{2g} (1 - \alpha_i t).$$

Then

$$\frac{Z_K'(t)}{Z_K(t)} = \frac{1}{1-t} + \frac{1}{1-qt} - \sum_{i=1}^{2g} \frac{\alpha_i}{1-\alpha_i t} = \sum_{n=0}^{\infty} b_K(n+1) t^n,$$

where $b_K(n) = a_{K_n}(1) = 1 + q^n - \sum_{i=1}^{2g} \alpha_i^n.$ □

Define $\mathscr{L}_K(t) := Z_K'(t)/Z_K(t)$. Then it follows that the Riemann Hypothesis is equivalent to an apparently weaker inequality:

Corollary 5.3.5. *With notation as above, the following statements are equivalent:*

1. $|\alpha_i| = q^{1/2}$ *for all i.*
2. *There exist constants C_0, C_1 such that $|a_{K_n}(1) - q^n| \le C_0 + C_1 q^{n/2}$ for almost all positive integers n.*
3. *The radius of convergence of* $\mathscr{L}_K(t) - \dfrac{q}{1-qt}$ *is at least* $q^{-1/2}$.

Proof. It is obvious that 1) implies 2). Assuming 2) and using $a_{K_n}(1) = b_K(n)$, we get

$$|\mathscr{L}_K(t) - \frac{q}{1-qt}| \le f(t) + \frac{C_0}{1-t} + \frac{C_1}{1-q^{1/2}t}$$

for some polynomial $f(t)$, and 3) follows. From (5.3.4) it is immediate that the radius of convergence of $\mathscr{L}_K(t) - \dfrac{q}{1-qt}$ is $\min_i |\alpha_i|^{-1}$. Thus, assuming 3) we have $|\alpha_i| \le q^{1/2}$ for all i. However, since $\alpha_i \alpha_{2g-i+1} = q$ by (5.2.5), we see that 3) implies 1). □

Our approach to proving the Riemann Hypothesis for K will be to count the points of K_n for all sufficiently large n and show that 2) holds above. We refer to points of K_n as "k_n-rational points." The key upper bound is provided by the Stöhr–Voloch theorem (4.4.24). Then a Galois-theoretic argument due to Bombieri [2] converts the upper bound to a lower bound.

Before proceeding further, we need to discuss the Frobenius map. Let $\overline{k}$ denote the algebraic closure of k. Recall that by (3.2.5), $\overline{K} := \overline{k} \otimes_k K$ is a field, whose full field of constants is obviously $\overline{k}$. By (3.2.6) the points of K_n can be identified with the points of $\overline{K}$ that are defined over k_n for any positive integer n.

Let $\mathfrak{f}_0 : K \to K$ be the qth-power map, where $|k| = q$, and let $\mathfrak{f} = \mathfrak{f}_K := 1 \otimes \mathfrak{f}_0 : \overline{K} \to \overline{K}$. The map $\mathfrak{f}$ is an isomorphism of $\overline{K}$ into itself which is called the *Frobenius map*. Note that because K is defined over k, $\mathfrak{f}$ is the identity on scalars. If we extend scalars to k_n, we have $\overline{K_n} = \overline{K}$, and the resulting Frobenius map is obviously just $\mathfrak{f}^n$. We let $\mathfrak{f}$ act on the points of $\overline{K}$ as defined in Section 3.5. Recall that if P is a point of $\overline{K}$ and $x \in \overline{K}$, then $x(P)$ denotes the residue class of $x \mod P$.

There is some subtlety involved in the definition of $\mathfrak{f}$. Note that there is an obvious action of $\operatorname{Gal}(\overline{k}/k)$ on $\overline{K}$ via $\sigma \mapsto \sigma \otimes 1$, which works over any field k. In particular, the usual Frobenius automorphism of $\overline{k}$ acts on $\overline{K}$ this way, but this is *not* the map $\mathfrak{f}$ defined above. Also note that $\mathfrak{f}$ is different from the qth-power map.

Lemma 5.3.6. *Let $\mathfrak{f} = \mathfrak{f}_K$ be the Frobenius map and let $Q \in \mathbb{P}_{\overline{K}}$. Then $Q^{\mathfrak{f}} = Q$ if and only if Q is defined over k.*

Proof. Let $x \in \mathscr{O}_Q \cap K$. By definition, we have $x(Q^{\mathfrak{f}}) = \mathfrak{f}(x)(Q) = x(Q)^q$. It follows that if $Q^{\mathfrak{f}} = Q$, then $x(Q) \in k$ for all $x \in \mathscr{O}_Q \cap K$. Putting $P := Q \cap K$, this means that $\deg P = 1$, and hence Q is defined over k. Conversely, suppose that P is a point of K. Since $\mathfrak{f}(x) = x^q$ for all $x \in K$, it follows that $\mathfrak{f}^{-1}(\mathscr{O}_P) = \mathscr{O}_P$, and thus $P^{\mathfrak{f}} = P$. Since P is a point, (3.2.6) implies that Q is the unique point of $\overline{K}$ containing P, and therefore $Q^{\mathfrak{f}} = Q$. □

We next consider a finite extension K'/k of K/k.[2] Then there is a natural inclusion $\overline{K} \subseteq \overline{K'}$. Since $\mathfrak{f}_{K'}$ is the identity on scalars and restricts to the qth-power map on K', it agrees with $\mathfrak{f}_K$ on $\overline{K}$.

Suppose, in addition, that K'/K is Galois. Then every automorphism $\sigma \in \operatorname{Gal}(K'/K)$ extends to an automorphism $1 \otimes \sigma$ of $\overline{K'} = \overline{k} \otimes_k K'$ that is the identity on $\overline{K}$. Since the $1 \otimes \sigma$ are all distinct, we have $|\overline{K'} : \overline{K}| \ge |K' : K|$. However, if $\{u_1, \ldots, u_n\}$ is a K-basis for K', then $\{1 \otimes u_1, \ldots, 1 \otimes u_n\}$ certainly spans $\overline{K'}/\overline{K}$, so that $|\overline{K'} : \overline{K}| = |K' : K|$. Thus, the map $\sigma \mapsto 1 \otimes \sigma$ is an isomorphism

[2] Recall that the notation K'/k means that k is the full field of constants of K'.

$\mathrm{Gal}(K'/K) \simeq \mathrm{Gal}(\overline{K'}/\overline{K})$. By abuse of notation, we will identify these two groups in this way. This in fact identifies all the groups $\mathrm{Gal}(K'_n/K_n)$.

Fix a positive integer n. Let Q be a point of $\overline{K'}$ dividing the point P of $\overline{K}$ such that P is defined over k_n. In general, Q may not be defined over k_n, so $Q^{\mathfrak{f}^n}$ will be another point dividing P. By (3.5.1) we have $Q^{\mathfrak{f}^n} = Q^{\sigma}$ for some $\sigma \in \mathrm{Gal}(\overline{K'}/\overline{K})$. We call σ the *Frobenius substitution* at the point Q. Note that σ depends on n.

We are going to count all the points of $\overline{K'}$ lying over some k_n-rational point of $\overline{K}$, counting separately those points with a given Frobenius substitution. Therefore, for function fields $K'/k \supseteq K/k$ with Frobenius map $\mathfrak{f}$ and an automorphism $\sigma \in G := \mathrm{Gal}(\overline{K'}/\overline{K})$ we define

$$\mathbb{P}_n(K'/K,\sigma) := \{Q \in \mathbb{P}_{\overline{K'}} \mid e(Q|Q\cap\overline{K}) = 1 \text{ and } Q^{\mathfrak{f}^n} = Q^{\sigma}\}.$$

Note that if $Q \in \mathbb{P}_n(K'/K,\sigma)$ for some σ, then $Q\cap\overline{K}$ is defined over k_n by (5.3.6). Since only finitely many points of $\overline{K}$ are defined over k_n, the sets $\mathbb{P}_n(K'/K,\sigma)$ are finite, disjoint, and their union is the set of all points $Q \in \mathbb{P}_{\overline{K'}}$ such that $Q \cap K$ is defined over k_n and is unramified in $\overline{K'}$.

Put $p_n(K'/K,\sigma) := |\mathbb{P}_n(K'/K,\sigma)|$. Since $\overline{k}$ is algebraically closed, the number of points $Q \in \mathbb{P}_{\overline{K'}}$ dividing a given P is $|K' : K|$ by (2.1.17). Moreover, the total number of points $P \in \mathbb{P}_{\overline{K}}$ that are ramified in K' is finite by (2.4.9), regardless of the extension of k over which they are defined. It follows that

$$|\sum_{\sigma\in G} p_n(K'/K,\sigma) - |K' : K| a_{K_n}(1)| \le C'$$

for some constant C', or in other words, we have proved

Lemma 5.3.7. *With the above notation, there exists a constant C independent of n such that*

$$|a_{K_n}(1) - \frac{1}{|G|}\sum_{\sigma\in G} p_n(K'/K,\sigma)| \le C. \quad \square$$

We are now ready for the main part of the proof.

Lemma 5.3.8. *With the above notation, suppose that*

$$|k| = q = p^{2r} > 4g_{K'}^4(g_{K'}-1)^2.$$

Then $p_m(K'/K,\sigma) \le 1 + q^m + 2g_{K'}q^{m/2}$ for all $\sigma \in G$ and all positive integers m.

Proof. Fix $\sigma \in G$ and a positive integer m, and put $\tau := \mathfrak{f}^m\sigma^{-1}$, where $\mathfrak{f} = \mathfrak{f}_K$ is the Frobenius map. Then $\mathbb{P}_m(K'/K,\sigma)$ is just the set of fixed points of τ on $\mathbb{P}_{\overline{K'}}$ that are unramified over $\overline{K}$. Note that $\tau(\overline{K'}) = (\overline{K'})^{q^m}$, from which it follows that every fixed point of τ is in fact a strong fixed point. It is also clear that $\tau(\phi)$ is not a scalar multiple of ϕ for any nontrivial projective map ϕ. Thus, the hypotheses of (4.4.24) are satisfied. Since $\deg(\tau) = q^m$, the result now follows by applying (4.4.25) with $n = p^{rm} = q^{m/2}$. $\square$

With (5.3.8) in hand, it is not difficult to complete the proof of the Riemann Hypothesis.

Theorem 5.3.9 (Weil). *Let K be a function field over a finite field k of order q, and let $L_K(t) = \prod_{i=1}^{2g}(1-\alpha_i t)$ be the numerator of its zeta function. Then $|\alpha_i| = q^{1/2}$ for all i.*

Proof. Choose $x \in K$ so that $K/k(x)$ is separable (see (2.4.6)). The extension $K/k(x)$ may not be normal, but there is a Galois extension $K'/k(x)$ with $K \subseteq K'$ by (A.0.12). The full field of constants k' of K' may be a finite extension of k, but using (5.3.1) to extend k to k' and change notation if necessary, we may assume that $k' = k$. We may further assume that q is large enough to satisfy the hypotheses of (5.3.8).

Let $G := \mathrm{Gal}(K'/k(x))$. Then K'/K is Galois with Galois group $H := G_K$ by (A.0.16). Let n be an arbitrary positive integer. Applying (5.3.7) to both extensions yields constants C and C_1 such that

$$|a_{k_n(x)}(1) - \frac{1}{|G|}\sum_{\sigma\in G} p_n(K'/k(x),\sigma)| \le C, \tag{5.3.10}$$

and

$$|a_{K_n}(1) - \frac{1}{|H|}\sum_{\sigma\in H} p_n(K'/K,\sigma) \le C_1. \tag{5.3.11}$$

Note that trivially $|a_{k_n(x)}(1)| = q^n + 1$. So (5.3.10) says that the average value of $p_n(K'/k(x),\sigma)$ over all σ is about q^n. Since each term of this average is less than q^n plus a small amount by (5.3.8), it follows that each term is in fact close to q^n. More precisely, put $d := |K' : k(x)| = |G|$ and $g := g_{K'}$. Then for each n and each σ, (5.3.10) yields

$$p_n(K'/k(x),\sigma) + (d-1)(q^n + 1 + 2gq^{n/2}) \ge \sum_{\tau\in G} p_n(K'/k(x),\tau) \ge d(q^n + 1 - C),$$

whence

$$p_n(K'/k(x),\sigma) \ge q^n - (d-1)(1+2gq^{n/2}) + d - dC.$$

It follows that

$$|p_n(K'/k(x)),\sigma) - q^n| \le A + Bq^{n/2}, \tag{5.3.12}$$

for constants A and B independent of n.

Note that for $\sigma \in H$, the sets $\mathbb{P}_n(K'/k(x),\sigma)$ and $\mathbb{P}_n(K'/K,\sigma)$ are essentially the same, differing by at most a finite number of points (independent of n) that are ramified over $\overline{k}(x)$ but not over $\overline{K}$. But now it follows from (5.3.11) and (5.3.12) that there exist constants A' and B' independent of n such that

$$|a_{K_n}(1) - q^n| \le A' + B'q^{n/2}$$

for all $n \ge 1$, and the theorem follows from (5.3.5). □

The above argument is essentially identical to [2] and also appears in [16]. For a variant, see [17]. The immediate corollary of (5.3.9) is the following important estimate for the number of points:

Corollary 5.3.13 (Weil). *Let K/k be a function field of genus g over a finite field k of order q. Then $|a_K(1) - q - 1| \leq 2gq^{1/2}$.* □

5.4 Exercises

Exercise 5.1. Compute the zeta function of the function field of the elliptic curve $y^2 + y = x + 1/x$ over $GF(2)$.

Exercise 5.2. Let $k := GF(3)$ be the field with three elements, and let $K := k(x,y)$, where $y^2 = 2(x^3 - x)^2 + 2$. Show that K has no k-rational points, but the image of the map $\{1,x,y\}$ has one k-rational point. Explain.

Exercise 5.3. Let k be the real numbers, and let $K := k(x,y)$, where $y^2 = -x^4 - (x-1)^4$. Show that every prime divisor of K has degree 2. Thus, K is a function field of genus 1 with no divisor of degree 1.

Exercise 5.4. Let $k := GF(q)$ and let $K := k(x,y)$, where $x^{q+1} + y^{q+1} = -1$ (see Exercise 4.5). Show that the Weil upper bound (5.3.13) for the number of points of K over $GF(q^2)$ is sharp.

Exercise 5.5. Let $k := GF(p^n)$ and K/k be a function field of genus g with $p \geq g \geq 3$. Let N be the number of points of K.

(i) Let $\phi : \mathbb{P}_K \longrightarrow \mathbb{P}^{g-1}$ be the canonical map, let $\mathfrak{f}$ be the Frobenius map, and let $j_1, \ldots, j_n$ be the $\mathfrak{f}$-orders of ϕ. Show that $j_n = g - 2$. [Hint: (4.4.9) and (4.4.22).]

(ii) Assume that ϕ is classical. Apply the Stöhr–Voloch theorem (4.4.24) to conclude that

$$N \leq 2q + g(g-1).$$

(iii) For what values of q and g is this bound better than the Weil upper bound (5.3.13)?

Exercise 5.6. Let $k := GF(q)$, where $q = p^m$, and let n be a positive integer. In this exercise we obtain a direct proof of the Riemann Hypothesis for the *Fermat curve*

$$V := V(X_0^n + X_1^n + X_2^n)$$

with function field $K = k(x,y)$, where $x^n + y^n = 1$. Let $E(n) := a_K(1) - q - 1$ be the error term. We will obtain a formula for $E(n)$ in terms of roots of unity.

(i) Show that $g_K = (n-1)(n-2)/2$. Thus, the Riemann Hypothesis asserts $|E(n)| \leq (n-1)(n-2)\sqrt{q}$.

(ii) Let $d := \gcd(m,n)$. Show that $E(n) = E(d)$. In particular, $E(n) = 0$ when m and n are relatively prime. For the remainder of the exercise, we assume $n \mid m$.

(iii) By a *character* of a finite abelian group G we will mean a homomorphism $\chi : G \to \mathbb{C}^\times$. The characters of G take values in the roots of unity and form a group under pointwise multiplication. We will repeatedly use below the fact that the nth roots of unity sum to zero for any $n > 1$. By convention, we extend characters of $k^\times$ to all of k via $\chi(0) := 0$.
Let $k^n = \{x^n \mid x \in k^\times\}$, and note that $0 \notin k^n$. Let $x \in k^\times$. Show that

$$\sum_{\chi^n=1} \chi(x) = \begin{cases} n & \text{if } x \in k^n, \\ 0 & \text{otherwise.} \end{cases}$$

Thus, we have

$$(*) \qquad \begin{aligned} |\{(x,y) \in k \times k \mid x^n + y^n = 1\}| &= 2n + \sum_{\substack{u+v=1 \\ uv \neq 0}} \sum_{\chi_1^n=1=\chi_2^n} \chi_1(u)\chi_2(v) \\ &= 2n + \sum_{\chi_1^n=1=\chi_2^n} J(\chi_1,\chi_2), \end{aligned}$$

where

$$J(\chi_1,\chi_2) := \sum_{x \in k} \chi_1(x)\chi_2(1-x).$$

(iv) Show that for $\chi \neq 1$ we have $J(\chi,1) = -1$.

(v) Let ξ be a primitive p^{th} root of unity and let τ be the character of the additive group k^+ defined by

$$\tau(x) = \xi^{\operatorname{tr}_{k/k_0}(x)},$$

where $k_0 := GF(p)$. For any multiplicative character χ, define

$$g(\chi) := \sum_{x \in k} \chi(x)\tau(x).$$

Show that

(**)
$$\begin{aligned} g(\chi_1)g(\chi_2) &= \sum_{\substack{u,v \\ v \neq 0}} \chi_1(u)\chi_2(v-u)\tau(v) + \sum_{u} \chi_1(u)\chi_2(-u) \\ &= \sum_{\substack{v,w \\ v \neq 0}} \chi_1(v)\chi_2(v)\chi_1(w)\chi_2(1-w)\tau(v) + \sum_{u} \chi_1(u)\chi_2(-u) \\ &= g(\chi_1\chi_2)J(\chi_1,\chi_2) + \begin{cases} \chi_2(-1)(q-1) & \text{if } \chi_1\chi_2 = 1, \\ 0 & \text{otherwise.} \end{cases} \end{aligned}$$

(vi) Show that $\overline{g(\chi)} = \chi(-1)g(\bar{\chi})$ for all χ. Conclude from (**) that $|g(\chi)| = \sqrt{q}$ for all $\chi \neq 1$, and that $|J(\chi_1,\chi_2)| = \sqrt{q}$ for all χ_1, χ_2 with $\chi_1\chi_2 \neq 1$.

(vii) Show that V has n points at infinity. Now use $(*)$ to obtain

$$E(n) = (n-1)(n-2) \sum_{\substack{\chi_1 \neq 1, \chi_2 \neq 1 \\ \chi_1\chi_2 \neq 1}} J(\chi_1, \chi_2),$$

and the desired bound follows from $(**)$.

Appendix A
Elementary Field Theory

Many, if not most, of the results in this chapter are standard in an elementary course on field theory, but we include them here anyway for the sake of completeness.

Recall that if $K \subseteq K'$ are fields, then K' is naturally a K-vector space, and we say that K'/K is a finite extension when $|K' : K| := \dim_K(K')$ is finite. Let K'/K be a finite extension with $u \in K'$, and let $M_{K'/K}(u)$ denote the K-linear transformation $K' \to K'$ defined by $x \mapsto ux$. Then we define the *trace* and *norm* via $\operatorname{tr}_{K'/K}(u) := \operatorname{tr} M_{K'/K}(u)$ and $N_{K'/K}(u) := \det M_{K'/K}(u)$. It is evident from this definition that $\operatorname{tr}_{K'/K}$ is a K-linear map from K' to K, and that $N_{K'/K}$ is a multiplicative homomorphism from K'^* to K^*.

Lemma A.0.1. *Suppose that K'/K is a finite extension of degree n, V is a K'-vector space of dimension m, and $A : V \to V$ is a K'-linear transformation. Then V has dimension mn over K, and*

$$\operatorname{tr}_K(A) = \operatorname{tr}_{K'/K}(\operatorname{tr}_{K'}(A)),$$
$$\det_K(A) = N_{K'/K}(\det_{K'}(A)).$$

Proof. Let $\{x_1, \ldots, x_n\}$ be a K-basis for K' and let $\{e_1, \ldots, e_m\}$ be a K'-basis for V. It is routine to verify that $\{x_i e_j \mid 1 \le i \le n, 1 \le j \le m\}$ is a K-basis for V. The matrix of A with respect to this basis has block form $M(a_{ik})$, where $Ae_i = \sum_k a_{ik} e_k$ for scalars $a_{ik} \in K'$ and $M(a)$ is the matrix of multiplication by a with respect to the basis $\{x_1, \ldots, x_n\}$.

The trace formula is now immediate. To get the determinant formula, reduce A to upper triangular block form by performing unimodular elementary row operations over K', which do not change the determinant over either field. Then reduce each block separately by unimodular row operations over K, and the determinant formula follows by inspection. □

Taking V above to be a finite extension K''/K', we have the following corollary:

Corollary A.0.2. *Suppose that $K \subseteq K' \subseteq K''$ are three fields, and $|K'' : K| < \infty$. Then*

$$|K'' : K| = |K'' : K'||K' : K|,$$

$$\mathrm{tr}_{K''/K} = \mathrm{tr}_{K'/K} \circ \mathrm{tr}_{K''/K'},$$

$$N_{K''/K} = N_{K'/K} \circ N_{K''/K'}. \quad \square$$

Corollary A.0.3. *Let $|K' : K| = n$, and let $u \in K'$ with minimum polynomial $f(X) := X^m + \sum_{i=0}^{m-1} a_i X^i$. Put $k := |K' : K(u)|$, then $\mathrm{tr}_{K'/K}(u) = ka_{m-1}$ and $N_{K'/K}(u) = a_0^k$.*

Proof. The point is that the characteristic polynomial of $M_{K(u)/K}(u)$ has degree $m = |K(u) : K|$, has coefficients in K, and is satisfied by u, so it is equal to $f(X)$. Hence $\det M_{K(u)/K}(u) = a_0$ and $\mathrm{tr} M_{K(u)/K}(u) = a_{m-1}$. It is trivial that $\mathrm{tr}_{K'/K(u)}(u) = ku$ and $N_{K'/K(u)} = u^k$. Using (A.0.2) we have $\mathrm{tr}_{K'/K}(u) = \mathrm{tr}_{K(u)/K}(ku) = ka_{m-1}$ and $N_{K'/K}(u) = N_{K(u)/K}(u^k) = a_0^k$. □

The following lemma is often useful for computing the trace.

Lemma A.0.4. *Let $|K' : K| = n$ and let $u \in K'$. If the roots of the characteristic polynomial of $M_{K'/K}(u)$ are $u = u_1, \ldots, u_n$ and $f(X)$ is a rational function with coefficients in K that is defined at each u_i, then*

$$\mathrm{tr}_{K'/K}(f(u)) = \sum_{i=1}^{n} f(u_i).$$

Proof. There exists an invertible matrix A with entries in some extension field of K such that $U := A^{-1} M_{K'/K}(u) A$ is upper triangular with $u_1, \ldots, u_n$ on the diagonal. Then $p(U) = A^{-1} M_{K'/K}(p(u)) A$ is an upper triangular matrix with $p(u_1), \ldots, p(u_n)$ on the diagonal, for any polynomial $p(X) \in K[X]$. Let $f(X) =: g(X)/h(X)$ where $g(X)$ and $h(X)$ are relatively prime polynomials. Our assumption is that $h(u_i) \neq 0$ for all i, whence $h(U)$ is invertible and $h(U)^{-1}$ is also upper triangular with $h(u_1)^{-1}, \ldots, h(u_n)^{-1}$ on the diagonal. We conclude that

$$A^{-1} M_{K'/K}(f(u)) A = g(U) h(U)^{-1}$$

is upper triangular with $f(u_1), \ldots, f(u_n)$ on the diagonal, and the result follows. □

The trace map turns out to be particularly useful because we can use it to define a bilinear form: Given a finite extension K'/K, define

$$(x, y)_{K'/K} := \mathrm{tr}_{K'/K}(xy).$$

We want to know when this form is nondegenerate. The answer involves the notion of separability.

Recall that an element u algebraic over K is *separable* over K if its minimum polynomial over K has distinct roots, and an extension $K \subseteq K'$ of fields is separable if every element of K' is separable over K.

Lemma A.0.5. *Let K be a field. An irreducible polynomial $f(X) \in K[X]$ has a repeated root iff $f(X) = g(X^p)$ for some irreducible polynomial $g(X)$, where $p = \text{char}(K)$.*

Proof. Over some extension field we have

$$f(X) = \prod_i (X - a_i),$$

hence

$$f'(X) = \sum_i \frac{f(X)}{X - a_i}.$$

If $f'(X) \neq 0$, it follows that a_i is a root of $f'(X)$ iff it is a repeated root of $f(X)$, and that $g(X) := \gcd(f(X), f'(X))$ will be nonconstant iff $f(X)$ has a repeated root. We conclude that the only way an irreducible polynomial can have a repeated root is for $f'(X)$ to be identically zero. This is easily seen to occur if and only if $f(X) = g(X^p)$ for some (necessarily irreducible) polynomial g, where $p = \text{char}(K)$. □

Notice that if every coefficient of a polynomial $g(X) \in k[X]$ is a pth power in k, then $g(X^p) = g_1(X)^p$ for some polynomial g_1, and therefore g is not irreducible. In general, the map $x \mapsto x^p$ is an isomorphism of K into itself. We say that K is *perfect* if this map is onto, or if $\text{char}(K) = 0$. Note that finite fields are perfect, because they are splitting fields of polynomials $X^{p^n} - X$, and algebraically closed fields are certainly perfect. The following corollary is immediate.

Corollary A.0.6. *Suppose that K is perfect. Then every irreducible polynomial over K has distinct roots.* □

If K' and L are extensions of K, we say that a map $K' \to L$ of fields is an embedding of K'/K into L if it restricts to the identity map on K.

Lemma A.0.7. *Let K' be an extension of K and let $\{\sigma_1, \dots, \sigma_n\}$ be distinct embeddings of K'/K into some extension of K. If $\{a_1, \dots, a_n\} \subseteq K$ with $\sum_i a_i \sigma_i(x) = 0$ for all $x \in K'$, then $a_i = 0$ for all i.*

Proof. Proceeding by way of contradiction, assume that there is a non-trivial dependence relation with notation chosen so that $a_i = 0$ for $i > m$ and $a_i \neq 0$ for $i \leq m$. We may further assume that m is minimal among all possible non-trivial dependence relations, and we note that $m \geq 2$. Choose $u \in K'$ with $\sigma_1(u) \neq \sigma_2(u)$. Then for all $x \in K'$ we have

$$0 = \sum_{i=1}^{m} a_i \sigma_i(ux) = \sum_{i=1}^{m} a_i \sigma_i(u) \sigma_i(x).$$

However, we also have

$$0 = \sigma_1(u) \sum_{i=1}^{m} a_i \sigma_i(x).$$

Subtracting these two relations gives a shorter nontrivial dependence relation on the σ_i, which is impossible. □

Theorem A.0.8. *Let K' be a finite extension of K. Then the following conditions are equivalent:*

1. *K' is separable over K.*
2. *$K' = K(u_1, \dots, u_n)$, where each u_i is separable over K.*
3. *The number of distinct embeddings of K'/K into a fixed algebraic closure $\overline{K}$ of K is equal to $|K' : K|$.*
4. *The trace form $(,)_{K'/K}$ is nondegenerate.*
5. *The trace $\mathrm{tr}_{K'/K}$ is nonzero.*

Proof. $1 \Rightarrow 2$: Trivial.

$2 \Rightarrow 3$: This is a standard argument, the point being that given a root u of an irreducible polynomial $f(X)$ in some extension K' of K, we can extend any embedding $\phi : K \to K'$ to a (unique) map $K[X] \to \overline{K}$ sending X to u. The kernel of this map is the ideal generated by f, so we have extended ϕ to an isomorphism ϕ_u : $K[X]/(f) \to K(u)$. Let $i : K \hookrightarrow K'$ be the inclusion map. Then for any embedding $\phi : K \to \overline{K}$ and any root $\bar{u}$ of f in $\overline{K}$ the map $\phi_{\bar{u}} \circ (i_u)^{-1} : K(u) \hookrightarrow \overline{K}$ is an extension of ϕ mapping u to $\bar{u}$. If f is separable, then there are $\deg f = |K(u) : K|$ distinct embeddings extending the identity map, because f has $\deg f$ distinct roots. If v is a root of some other separable irreducible polynomial over K, then v is also separable over $K(u)$, so we get $|K(u,v) : K(u)|$ distinct extensions of each of the embeddings of $K(u)$. By an obvious induction argument, we have $|K' : K|$ distinct embeddings of K'/K into $\overline{K}$.

$3 \Rightarrow 4$: Let $\{u_1, \dots, u_n\}$ be a K-basis for K', and let $\{\sigma_1, \dots, \sigma_n\}$ be the distinct embeddings of K'/K into $\overline{K}$. We claim that the matrix $D := (\sigma_i(u_j))$ is nonsingular. If not, there exist elements $a_i \in \overline{K}$, not all zero, such that the K-linear transformation $\sum_i a_i \sigma_i$ vanishes at u_j for all j and therefore vanishes identically on K', contrary to (A.0.7). We conclude that D is nonsingular, hence so is $D^t D = \mathrm{tr}_{K'/K}(u_i u_j)$, as required.

$4 \Rightarrow 5$: Trivial.

$5 \Rightarrow 1$: If K' is not separable over K, then there exists an element $u \in K' \setminus K$ whose minimum polynomial, $f(X)$, has a repeated root. By (A.0.5), $\mathrm{char}(K) = p$ and $f(X) = g(X^p)$ for some irreducible polynomial $g(X)$. Let $v = u^p$ and $E = K(v)$. Then u has minimum polynomial $X^p - v$ over E. Calculating with respect to the basis $\{1, u, u^2, \dots, u^{p-1}\}$ and noting that $\mathrm{tr}_{E(u)/E}(1) = p = 0$, it is easy to check that $\mathrm{tr}_{E(u)/E}(u^i) = 0$ for all i, whence $\mathrm{tr}_{E(u)/E} \equiv 0$. By repeated application of (A.0.2), $\mathrm{tr}_{K'/K}$ factors through $\mathrm{tr}_{E(u)/E}$ and is therefore zero. □

We say that K'/K is *purely inseparable* if $\mathrm{char}(K) = p > 0$ and for every $u \in K'$ we have $u^q \in K$ for some power q of p. In this case, u is a root of $X^q - a$ for some $a \in K$, which factors over $K(u)$ as $(X - u)^q$, so u is the only root of its minimum polynomial.

Corollary A.0.9. *Let K'/K be finite. Then the set of all elements of K' separable over K form a subfield K'_s that is separable over K, and the extension K'/K'_s is purely inseparable.*

Proof. Since the subfield of K' generated over K by any finite set of separable elements is separable over K by (A.0.8), the finiteness of $|K' : K|$ implies that there is a maximal separable extension K'_s/K consisting of all elements of K' separable over K.

If $u \in K' \setminus K'_s$ with minimum polynomial $f(X)$ over K'_s, then (A.0.5) yields $f(X) = g(X^p)$ for some irreducible polynomial $g(X)$, which evidently is the minimum polynomial of $v := u^p$. If g is not linear, we may continue in this way, eventually obtaining $f(X) = X^q - a$ for some power q of p and some element $a \in K'_s$. □

Corollary A.0.10. *Suppose that K_1 and K_2 are subfields of K' with $K := K_1 \cap K_2$ and $K' = K_1K_2$. Assume further that K_1/K is finite and separable and K_2/K is finite and purely inseparable. Then the natural map $K_1 \otimes_K K_2 \to K'$ is an isomorphism.*

Proof. The natural map is surjective because $K' = K_1K_2$. To show that it is injective, we proceed by way of contradiction, assuming that there are nonzero elements $x_i \in K_1$, $y_i \in K_2$ with

$$\sum_{i=1}^{n} x_i y_i = 0,$$

and that we have chosen such a relation with n minimal. Then the x_i and y_j are separately linearly independent over K, or else n would not be minimal. There is a power q of $p := \mathrm{char}(K)$ with $y_i^q \in K$ for all i. This implies that the x_i^q are linearly dependent over K, since the map $x \mapsto x^q$ is a homomorphism.

On the other hand, we have $\det(x_i^q, x_j^q) = \det(x_i, x_j)^q \neq 0$, where (u, v) is the trace form on K_1/K. This implies that the x_i^q are linearly independent over K. □

More generally, two subfields K_1 and K_2 of a field K' whose intersection contains K are said to be *linearly disjoint* over K if the natural map $K_1 \otimes_K K_2 \to K'$ is injective. Let $\{x_i \mid i \in I\}$ and $\{y_j \mid j \in J\}$ be (possibly infinite) K-bases for K_1 and K_2, respectively. Then $\{x_i \otimes y_j \mid i \in I, j \in J\}$ is a K-basis for $K_1 \otimes_K K_2$ by standard properties of the tensor product. It follows that K_1 and K_2 are linearly disjoint if and only if $\{x_i y_j \mid i \in I, j \in J\}$ is linearly independent over K, but this occurs if and only if $\{x_i \mid i \in I\}$ is linearly independent over K_2.

Suppose that K_1 and K_2 are linearly disjoint over K and that I is finite. Then the image of the natural map $R := K_1K_2$ is an integral domain that is finite-dimensional over K_2. Since $K_2[X]$ is a principal ideal domain, it follows that $K_2[x]$ is a field for all $x \in R$, and therefore R is a field.

Now suppose, in addition, that E is an intermediate field $K \subseteq E \subseteq K_2$. Let $\{u_l \mid l \in L\}$ be a K-basis for E and let $\{v_m \mid m \in M\}$ be an E-basis for K_2. Then $\{u_l v_m \mid l \in L, m \in M\}$ is a K-basis for K_2. It follows that $\{x_i u_l \mid i \in I, l \in L\}$ is linearly independent, and therefore K_1E is a field. Moreover, $\{x_i v_m \mid i \in I, m \in M\}$ is linearly independent over E. We have proved

Lemma A.0.11. *Suppose that K_1 and K_2 are linearly disjoint over K, and $|K_1 : K|$ is finite. Then $K_1 \otimes_K K_2$ is a field and*

$$|K_1 \otimes_K K_2 : K_2| = |K_1 : K|.$$

If E is an intermediate field $K \subseteq E \subseteq K_2$, then K_1 and E are linearly disjoint over K, and K_1E and K_2 are linearly disjoint over E. □

We call an extension K'/K *normal* if $K' = K(u_1, \ldots, u_n)$, where the u_i are all the roots of some polynomial $f \in K[X]$. If $\bar{K} \supseteq K'$ is any algebraic closure of K, the u_i are evidently permuted by all embeddings of K'/K into $\bar{K}$, which therefore induce automorphisms of K'/K. We denote by $\mathrm{Gal}(K'/K)$ the group of automorphisms of K' fixing K elementwise. If K'/K is both normal and separable, we call it a *Galois extension* of K.

Corollary A.0.12. *Every finite separable extension is contained in a Galois extension.*

Proof. Let $K' = K(u_1, \ldots, u_n)$, where the u_i are separable over K. Adjoin the remaining roots, if any, of the minimum polynomial of each u_i to K' and apply (A.0.8). □

Corollary A.0.13. *A finite extension K'/K is Galois if and only if $|K' : K| = |\mathrm{Gal}(K'/K)|$.*

Proof. Put $G := \mathrm{Gal}(K'/K)$. If K'/K is Galois, then there are $|K' : K|$ distinct embeddings of K'/K into some algebraic closure of K' by (A.0.8). As discussed above, these embeddings stabilize K', and thus we get $|G| = |K' : K|$.

Conversely, (A.0.8) implies that K'/K is separable. Let $K' = K(u_1, \ldots, u_n)$ and define

$$f(X) := \prod_{i=1}^{n} \prod_{\sigma \in G} (X - \sigma(u_i)).$$

The coefficients of $f(X)$ are G-invariant, and therefore $f \in K[X]$. Since all roots of f lie in K' and generate K'/K, we conclude that K'/K is normal. □

Lemma A.0.14. *Let V be a vector space over an infinite field, and suppose that $W_1, \ldots, W_m$ are proper subspaces. Then V has a basis which is disjoint from any of the W_i. In particular, V is not the union of the W_i.*

Proof. We proceed by induction on m, the result being vacuously true for $m = 0$. Assume, then, that $\{v_1, \dots, v_n\}$ is a basis such that

$$v_i \notin W_j \text{ for } 1 \le i \le n \text{ and } 1 \le j < m.$$

If none of the v_i lie in W_m, we are done. Otherwise, choose notation so that $v_i \in W_m$ if and only if $1 \le i \le r$. Since W_m is a proper subspace, we have $r < n$. Fix $i \le r$, and consider the set of vectors

$$\{u_\alpha := v_n + \alpha v_i \mid \alpha \in k\}.$$

None of the u_α lie in W_n, because $v_n \notin W_m$. If $\{u_\alpha, u_\beta\} \subseteq W_j$ for some $j < m$ and some $\alpha \neq \beta$, we get

$$v_i = \frac{u_\alpha - u_\beta}{\alpha - \beta} \in W_j.$$

But then $v_n = u_\alpha - \alpha v_i \in W_j$, which is not the case. So there is at most one u_α in W_j for each j, and therefore we can choose $\alpha_i \in k$ such that $u_i := u_{\alpha_i} \notin W_j$ for any j, because k is infinite. The desired basis is then $\{u_1, \dots, u_r, v_{r+1}, \dots, v_n\}$. □

Corollary A.0.15. *Suppose that K' is a finite extension of K such that there are only finitely many intermediate fields between K and K'. Then $K' = K(u)$ for some element $u \in K'$.*

Proof. If K is a finite field, then so is K'. Since there are at most n roots of the polynomial $X^n - 1$ in K' for any n, the multiplicative group of nonzero elements of K' must be cyclic by the fundamental theorem of abelian groups. Taking u to be a generator, we have $K' = K(u)$. If K is infinite, there is an element $u \in K'$ that does not lie in any proper subfield by (A.0.14), and thus $K' = K(u)$. □

Theorem A.0.16 (Fundamental Theorem of Galois Theory). *Let K'/K be a Galois extension with $G := \mathrm{Gal}(K'/K)$. For any intermediate field $K \subseteq E \subseteq K'$, let $G_E := \{g \in G \mid g(u) = u \text{ for all } u \in E\}$. Then K'/E is Galois with $\mathrm{Gal}(K'/E) = G_E$, and the map $E \to G_E$ is a one-to-one inclusion-reversing correspondence between subfields of K' containing K and subgroups of G. Moreover, E/K is normal if and only if G_E is a normal subgroup of G, in which case restriction induces a natural isomorphism $G/G_E \simeq \mathrm{Gal}(E/K)$.*

Proof. If K' is normal (resp. separable) over K, it is also normal (resp. separable) over any intermediate field E. Hence K'/E is Galois, and there is an inclusion $\mathrm{Gal}(K'/E) \subseteq G$ with $G_E = \mathrm{Gal}(K'/E)$. Moreover, the map $E \to G_E$ is clearly inclusion-reversing. Given a subgroup $H \subseteq G$, let E_H be the subfield of K' elementwise fixed by H. Then $H \subseteq G_E$. Since $E_{G_E} \supseteq E$, we get $E_{G_E} = E$ by (A.0.13). Thus, the map $E \to G_E$ is one-to-one.

In particular, there are only finitely many intermediate fields between E_H and K' for any subgroup $H \subseteq G$. By (A.0.15), $K' = E_H(u)$ for some $u \in K'$. However, the polynomial

$$\prod_{\sigma \in H} (X - \sigma(u))$$

has degree $|H|$ and coeficients in E_H, whence $|K' : E_H| \leq |H|$. Since $H \subseteq G_{E_H}$, we get $H = G_{E_H}$ by (A.0.13). Thus, the map $E \to G_E$ is a one-to-one correspondence, as asserted.

If E/K is normal, restriction yields a natural map $G \to \mathrm{Gal}(E/K)$ whose kernel is G_E. Since the image of this map has order

$$\frac{|G|}{|G_E|} = \frac{|K' : K|}{|K' : E|} = |E : K|,$$

it induces a natural isomorphism $G/G_E \simeq \mathrm{Gal}(E/K)$ by (A.0.13).

Conversely, suppose that H is normal in G, and $u \in E_H$. Then for any $g \in G$ and $h \in H$ we have $hg(u) = gg^{-1}hg(u) = gh_1(u) = g(u)$, where $h_1 := g^{-1}hg$. This means that $g(u) \in E_H$, and it follows immediately that E_H/K is normal. □

Corollary A.0.17. *Suppose that K' is a finite separable extension of K. Then $K' = K(u)$ for some $u \in K'$.*

Proof. Since K' is contained in a Galois extension K'' of K by (A.0.12), there are only finitely many intermediate fields between K' and K by (A.0.16), and the result follows from (A.0.15). □

When K'/K is Galois, the trace and norm have particularly nice expressions:

Lemma A.0.18. *Let K'/K be a Galois extension with $G := \mathrm{Gal}(K'/K)$, and let $u \in K'$. Then*

$$\mathrm{tr}_{K'/K}(u) = \sum_{\sigma \in G} \sigma(u),$$
$$N_{K'/K}(u) = \prod_{\sigma \in G} \sigma(u).$$

Proof. Let $H := G_{K(u)} = \mathrm{Gal}(K'/K(u))$ and let $\{x_1, \ldots, x_m\}$ be a set of coset representatives for H in G. Then the set $\{x_1(u), \ldots, x_m(u)\}$ is the set of distinct G-conjugates of u and is therefore the set of distinct roots of the minimum polynomial $f(X)$ of u over K. Put $f(X) = \sum_{i=0}^m a_i X^i$, and consider the polynomial

$$\prod_{\sigma \in G} (X - \sigma(u)) = \prod_{i=1}^m \prod_{\sigma \in H} (X - x_i \sigma(u)) = \prod_{i=1}^m (X - x_i(u))^h = f(X)^h,$$

where $h = |H| = |K' : K(u)|$. It follows that

$$\sum_{\sigma \in G} \sigma(u) = h a_{m-1}, \quad \text{and} \quad \prod_{\sigma \in G} \sigma(u) = a_0^h.$$

The result now follows from (A.0.3). □

The case that K is a finite field deserves special mention. The point here is that K must have characteristic $p > 0$, so it is a finite extension of the prime field F_p of order p. In particular, $|K| = p^n$ for some integer n, so the multiplicative group of K has order $p^n - 1$. It follows that every nonzero element of K satisfies the polynomial $X^{p^n - 1} - 1$, and hence K is the splitting field of $X^{p^n} - X$ over F_p.

Conversely, the polynomial $X^{p^m} - X$ is separable for all m. If u and v are roots, then so are $u+v$ and uv since we are in characteristic p. Thus, the roots form a field of order p^m, so there is a unique finite field F_q of order $q = p^m$ for any $m > 0$. All of these fields are contained in $\overline{F_p}$, the algebraic closure of F_p. Indeed, F_{p^n} is just the subfield of $\overline{F_p}$ fixed by ϕ^n, where $\phi(x) := x^p$ is the *Frobenius* automorphism of $\overline{F_p}$. In particular, if $n \mid m$, then $F_{p^n} \subseteq F_{p^m}$. Conversely, if $|F_{q'} : F_q| = r$, then $q' = q^r$. Summarizing all of this, we have

Theorem A.0.19. *Let p be a prime integer and n any nonnegative integer. There exists a unique finite field F_q of order $q = p^n$, and it is the splitting field of $X^{p^n} - X$ over the prime field F_p of order p. We have $F_{p^n} \subseteq F_{p^m}$ if and only if $n \mid m$, in which case the extension is Galois with cyclic Galois group of order m/n generated by the nth power of the Frobenius map: $\sigma(x) := x^{p^n}$.* □

We turn now from the algebraic case to the transcendental case. Let A be a k-algebra. We say that $a_1, a_2, \ldots, a_n \in A$ are *algebraically dependent* (over k) if there exists a nonzero polynomial $f \in k[X_1, \ldots, X_n]$ such that $f(a_1, \ldots, a_n) = 0$, and *algebraically independent* otherwise. In particular, an element a is transcendental over k iff $\{a\}$ is algebraically independent.

Now suppose that K is an extension field of k. For any subset $S \subseteq K$, let $\overline{k(S)}$ be the subfield of K consisting of all elements of K algebraic over $k(S)$. We say that S satisfies a minimal dependence relation if S is algebraically dependent, but every proper subset of S is algebraically independent. In this case, if $|S| = n+1$, we have a polynomial $f \in k[X_0, X_1, \ldots, X_n]$ with $f(s_0, s_1, \ldots, s_n) = 0$. If m is the highest power of X_0 that appears in any monomial of f, we can write

$$f(s_0, s_1, \ldots, s_n) = \sum_{j=0}^{m} g_j(s_1, s_2, \ldots, s_n) s_0^i = 0,$$

where $g_j \in k[X_1, \ldots, X_n]$. Since $\{s_1, s_2, \ldots, s_n\}$ is algebraically independent, the above identity specializes to a nonzero polynomial over $k(s_1, \ldots, s_n)$ satisfied by s_0. The same argument holds for any s_i, so we have proved

Lemma A.0.20. *Suppose that $S \subseteq K$ satisfies a minimal dependence relation. Then every element of S is algebraic over the subfield of K generated by the other elements of S.*

We will say that a subset $T \subseteq K$ *spans* K/k if $K = \overline{k(T)}$.

Theorem A.0.21.

1. *Every minimal spanning subset of K/k is algebraically independent over k.*

2. *Every maximal algebraically independent subset of K/k spans K/k.*

3. *If T spans K/k for some finite subset $T \subseteq K$, and $S \subseteq K$ is algebraically independent over k, then $|S| \leq |T|$.*

Proof. 1. If $B \subseteq K$ is a spanning set that is not independent over k, then there exists a minimal dependence relation among some (finite) subset $B' \subseteq B$. If $b \in B'$, then $B \setminus \{b\}$ spans K/k by (A.0.20).

2. Suppose $B \subseteq K$ is a maximal algebraically independent set over k, and let $x \in K$. Then $B' := B \cup \{x\}$ is algebraically dependent. Any minimal k-dependence relation among the elements of B' must involve x because B is algebraically independent. Then $x \in \overline{k(B)}$ by (A.0.20), and thus $\overline{k(B)} = K$.

3. Choose $R \subseteq T \cup S$ such that $|R| = |T|$, $\overline{k(R)} = K$, and $|R \cap S|$ is maximal with these properties. We claim that $S \subseteq R$. If not, choose $s \in S \setminus R$. Then since $s \in \overline{k(R)}$, there is some minimal dependence relation $g(s, r_1, \ldots, r_m) = 0$ for some $m \leq |R|$. However, notation can be chosen so that $r_1 \notin S$ because S is algebraically independent. Then $r_1 \in \overline{k(s, r_2, \ldots, r_m)}$ by (A.0.20). Put $R' = R \setminus \{r_1\} \cup \{s\}$. Then $|R'| = |R| = |T|$, $\overline{R'} = \overline{R} = K$, and $|R' \cap S| > |R \cap S|$, contradicting our choice of R. It follows that $S \subseteq R$ and hence $|S| \leq |T|$, as required. □

Corollary A.0.22. *If K is a finitely generated extension of k, then every maximal algebraically independent subset over k has the same cardinality.* □

We call any such subset a *transcendence basis* and we call its cardinality the *transcendence degree* of K over k, denoted by $\operatorname{trdeg}(K/k)$.

Corollary A.0.23. *Suppose that R is an integral domain whose field of fractions K has finite transcendence degree over some subfield k. Then R contains a transcendence basis for K over k.*

Proof. Let $\{x_1, x_2, \ldots, x_n\}$ be a transcendence basis for K over k. For each i there are elements $r_i, s_i \in R$ with $x_i = r_i/s_i$. Then K is spanned by $\{r_1, \ldots, r_n, s_1, \ldots, s_n\}$. By (A.0.21) this set contains a basis. □

Accordingly, we define the transcendence degree of a k-algebra R to be the transcendence degree of its field of fractions.

References

[1] E. Arbarello, M. Cornalba, P. Griffiths, J. Harris, *Geometry of Algebraic Curves*, Vol. I, Springer-Verlag 1985.

[2] E. Bombieri, *Counting Points over Finite Fields*, Sem. Bourbaki **430** (1972–73) pp. 234–241.

[3] N. Bourbaki, *Commutative Algebra*, Hermann 1972.

[4] C. Chevalley, *Introduction to the Theory of Algebraic Functions of One Variable*, American Mathematical Society 1951.

[5] M. Deuring, *Lectures on the Theory of Algebraic Functions of One Variable*, Springer-Verlag 1973.

[6] M. Eichler, *Introduction to the Theory of Algebraic Numbers and Algebraic Functions*, Academic Press 1966.

[7] D. Eisenbud, *Commutative Algebra with a View Toward Algebraic Geometry*, Springer-Verlag 1995.

[8] W. Fulton, *Intersection Theory*, Springer-Verlag 1984.

[9] D. Gorenstein, *An Arithmetic Theory of Adjoint Plane Curves*, Trans. Amer. Math. Soc. **72** (1952) pp. 414–436.

[10] R. Hartshorne, *Algebraic Geometry*, Springer-Verlag 1977.

[11] E. Hecke, *Lectures on the Theory of Algebraic Numbers* (translation) Springer-Verlag 1981.

[12] S. DiPippo and E. Howe, *Real Polynomials with all Roots on the Unit Circle and Abelian Varieties over Finite Fields*, J. Number Theory **73** (1998) pp. 433–434.

[13] K. Iwasawa, *Algebraic Functions*, Translations of Mathematical Monographs Vol. 118, American Mathematical Society, 1972.

[14] S. Lang, *Introduction to Algebraic and Abelian Functions*, Springer-Verlag 1982.

[15] R. Miranda, *Algebraic Curves and Riemann Surfaces*, American Mathematical Society 1995.

[16] C. Moreno, *Algebraic Curves over Finite Fields*, Cambridge University Press 1991.

[17] H. Stichtenoth, *Algebraic Function Fields*, Springer-Verlag 1993.

[18] K. Stöhr, *On Singular Primes in Function Fields*, Arch. Math **50** (1988) pp. 156–163.

[19] K. Stöhr and J. Voloch, *Weierstrauss Points and Curves over Finite Fields*, Proc. London Math. Soc. (3), 52 (1986), 1–19.

[20] J. Tate, *Residues of Differentials on Curves*, Ann. Scient. Ec. Norm. Sup. (1968) pp. 149–159.

[21] J. Tate, *Genus Change in Inseparable Extensions of Function Fields*, Proc. Amer. Math. Soc. (3) 1952, pp. 400–406.

[22] B.L. Van der Waerden, *Modern Algebra, Vol. I*, Ungar 1953.

[23] A. Weil, *Sur les courbes algébriques et les variétés qui s'en déduisent*, Hermann 1948.

Index

absolutely irreducible, 54
adele, 43–44, 47
  class group, 44
adjoint
  conditions, 143
  curves, 145
  divisor, 141
  function, 143

Bézout's theorem, 113
Bombieri, Enrico, 151

canonical map, 121
class number, 152
completion, 16–18
  of a graded algebra, 26
  of a module, 18
conductor
  local, 80
  of a plane curve, 144
conorm, 69

degree, 113
  of plane curve, 137
derivation, 24
  extension of, 26
  generalized, 27
different, 75–82
  exponent, 76, 80
  inverse, 81
  of an element, 78
  transitivity of, 76
differential form, 24, 61
  divisor of, 61
  exact, 25
  module of, 24
  regular, 62
discrete valuation ring, 5
  completion of, 20–23
divisor, 40
  canonical, 48
  class group, 42, 153
    order of, 152
  degree of, 41
  field of definition, 75
  group, 41
  linear equivalence, 42
  partial ordering on, 42
  prime, 40
    singular, 73
  principal, 41
  separable prime, 57
  special, 47, 129
  splitting field for, 75

dual
  map, 139
  plane, 136

elliptic curve, 52, 122
  group of, 53
extension
  Galois, 89–93, 169–171
  Kummer, 101
  purely inseparable, 168
  scalar, 72
  separable, 26, 55, 57, 166–168
  tamely ramified, 77
  totally ramified, 13, 78, 91
  transcendental, 172
  unramified, 12, 101
  weakly separable, 58
  wildly ramified, 77, 93

finite fields, 172
finitepotent
  map, 30
  subspace, 31
formal power series, 21–23
Frobenius
  map, 158
  substitution, 159
function field, 40
  geometric, 54
  singular, 88

generic
  coordinates, 142
  line, 141
  map, 142
genus, 47
  of a plane curve, 143
Goppa codes, 65
Gorenstein relations, 146
graded
  algebra, 25, 104
  ideal, 104
  module, 104

Hasse derivative, 28, 64
  chain rule, 29
Hensel's lemma, 19
Hilbert
  basis theorem, 116
  nullstellensatz, 105
hyperplane at infinity, 111

integral closure, 3, 4
intersection
  divisor, 112
  multiplicity, 112
    of plane curves, 137

Jacobian, 51, 70

Laurent series, 22, 64
linearly disjoint, 168
local parameter, 5

Nakayama's lemma, 3
Newton's algorithm, 19
norm, 69, 70, 164, 171
  absolute, 150

osculating plane, 129

plane model, 110
primitive element, 171
product formula, 45
projective
  curve, 107
  embedding, 120, 122
  space, 103
  variety, 106
projective map, 108–114
  birational, 110
  canonical, 121
  divisor of, 110
  effective, 109
  natural, 110
  normalized, 112
  singularity of, 114

ramification
  tame, 77
  wild, 77
ramification index, 10, 14, 57
rational function, 107
residue, 34–36
  form, 33
  local, 64
  theorem, 60
  trace formula, 35, 72

residue degree, 11, 14
residue field
  separable, 58
Riemann Hypothesis, 160
Riemann zeta function, 150
Riemann–Hurwitz formula, 76
Riemann–Roch theorem, 50–51, 62, 122, 129, 134, 146, 153, 155

separating variable, 26, 27, 55
singularities, 114–122
  of a plane curve, 120, 143, 146
Smith normal form, 6
splitting field, 73
Stöhr–Voloch theorem, 133, 158
strong approximation theorem, 51, 120, 147
strong fixed point, 130

tangent line, 53, 138
Tate, John, 30, 59
Taylor's theorem, 64
trace, 79, 164–168, 171
  of a differential form, 71
transcendence degree, 173

ultrametric inequality, 1

valuation extension theorem, 3, 109
valuation ring, 1

weak approximation theorem, 7, 71, 78, 80, 82, 89
Weierstrass
  divisor, 126, 135
  gap sequence, 129
  points, 127
Weil
  André, 151, 160
  bound, 161
  differential, 48, 60
Wronskian, 124

Zariski closed set, 103
zeta function, 150

Graduate Texts in Mathematics

(continued from page ii)

64 EDWARDS. Fourier Series. Vol. I. 2nd ed.
65 WELLS. Differential Analysis on Complex Manifolds. 2nd ed.
66 WATERHOUSE. Introduction to Affine Group Schemes.
67 SERRE. Local Fields.
68 WEIDMANN. Linear Operators in Hilbert Spaces.
69 LANG. Cyclotomic Fields II.
70 MASSEY. Singular Homology Theory.
71 FARKAS/KRA. Riemann Surfaces. 2nd ed.
72 STILLWELL. Classical Topology and Combinatorial Group Theory. 2nd ed.
73 HUNGERFORD. Algebra.
74 DAVENPORT. Multiplicative Number Theory. 3rd ed.
75 HOCHSCHILD. Basic Theory of Algebraic Groups and Lie Algebras.
76 IITAKA. Algebraic Geometry.
77 HECKE. Lectures on the Theory of Algebraic Numbers.
78 BURRIS/SANKAPPANAVAR. A Course in Universal Algebra.
79 WALTERS. An Introduction to Ergodic Theory.
80 ROBINSON. A Course in the Theory of Groups. 2nd ed.
81 FORSTER. Lectures on Riemann Surfaces.
82 BOTT/TU. Differential Forms in Algebraic Topology.
83 WASHINGTON. Introduction to Cyclotomic Fields. 2nd ed.
84 IRELAND/ROSEN. A Classical Introduction to Modern Number Theory. 2nd ed.
85 EDWARDS. Fourier Series. Vol. II. 2nd ed.
86 VAN LINT. Introduction to Coding Theory. 2nd ed.
87 BROWN. Cohomology of Groups.
88 PIERCE. Associative Algebras.
89 LANG. Introduction to Algebraic and Abelian Functions. 2nd ed.
90 BRØNDSTED. An Introduction to Convex Polytopes.
91 BEARDON. On the Geometry of Discrete Groups.
92 DIESTEL. Sequences and Series in Banach Spaces.
93 DUBROVIN/FOMENKO/NOVIKOV. Modern Geometry—Methods and Applications. Part I. 2nd ed.
94 WARNER. Foundations of Differentiable Manifolds and Lie Groups.
95 SHIRYAEV. Probability. 2nd ed.
96 CONWAY. A Course in Functional Analysis. 2nd ed.
97 KOBLITZ. Introduction to Elliptic Curves and Modular Forms. 2nd ed.
98 BRÖCKER/TOM DIECK. Representations of Compact Lie Groups.
99 GROVE/BENSON. Finite Reflection Groups. 2nd ed.
100 BERG/CHRISTENSEN/RESSEL. Harmonic Analysis on Semigroups: Theory of Positive Definite and Related Functions.
101 EDWARDS. Galois Theory.
102 VARADARAJAN. Lie Groups, Lie Algebras and Their Representations.
103 LANG. Complex Analysis. 3rd ed.
104 DUBROVIN/FOMENKO/NOVIKOV. Modern Geometry—Methods and Applications. Part II.
105 LANG. $SL_2(\mathbf{R})$.
106 SILVERMAN. The Arithmetic of Elliptic Curves.
107 OLVER. Applications of Lie Groups to Differential Equations. 2nd ed.
108 RANGE. Holomorphic Functions and Integral Representations in Several Complex Variables.
109 LEHTO. Univalent Functions and Teichmüller Spaces.
110 LANG. Algebraic Number Theory.
111 HUSEMÖLLER. Elliptic Curves.
112 LANG. Elliptic Functions.
113 KARATZAS/SHREVE. Brownian Motion and Stochastic Calculus. 2nd ed.
114 KOBLITZ. A Course in Number Theory and Cryptography. 2nd ed.
115 BERGER/GOSTIAUX. Differential Geometry: Manifolds, Curves, and Surfaces.
116 KELLEY/SRINIVASAN. Measure and Integral. Vol. I.
117 J.-P. SERRE. Algebraic Groups and Class Fields.
118 PEDERSEN. Analysis Now.
119 ROTMAN. An Introduction to Algebraic Topology.
120 ZIEMER. Weakly Differentiable Functions: Sobolev Spaces and Functions of Bounded Variation.
121 LANG. Cyclotomic Fields I and II. Combined 2nd ed.
122 REMMERT. Theory of Complex Functions. *Readings in Mathematics*
123 EBBINGHAUS/HERMES et al. Numbers. *Readings in Mathematics*

124 DUBROVIN/FOMENKO/NOVIKOV. Modern Geometry—Methods and Applications. Part III
125 BERENSTEIN/GAY. Complex Variables: An Introduction.
126 BOREL. Linear Algebraic Groups. 2nd ed.
127 MASSEY. A Basic Course in Algebraic Topology.
128 RAUCH. Partial Differential Equations.
129 FULTON/HARRIS. Representation Theory: A First Course.
Readings in Mathematics
130 DODSON/POSTON. Tensor Geometry.
131 LAM. A First Course in Noncommutative Rings. 2nd ed.
132 BEARDON. Iteration of Rational Functions.
133 HARRIS. Algebraic Geometry: A First Course.
134 ROMAN. Coding and Information Theory.
135 ROMAN. Advanced Linear Algebra.
136 ADKINS/WEINTRAUB. Algebra: An Approach via Module Theory.
137 AXLER/BOURDON/RAMEY. Harmonic Function Theory. 2nd ed.
138 COHEN. A Course in Computational Algebraic Number Theory.
139 BREDON. Topology and Geometry.
140 AUBIN. Optima and Equilibria. An Introduction to Nonlinear Analysis.
141 BECKER/WEISPFENNING/KREDEL. Gröbner Bases. A Computational Approach to Commutative Algebra.
142 LANG. Real and Functional Analysis. 3rd ed.
143 DOOB. Measure Theory.
144 DENNIS/FARB. Noncommutative Algebra.
145 VICK. Homology Theory. An Introduction to Algebraic Topology. 2nd ed.
146 BRIDGES. Computability: A Mathematical Sketchbook.
147 ROSENBERG. Algebraic K-Theory and Its Applications.
148 ROTMAN. An Introduction to the Theory of Groups. 4th ed.
149 RATCLIFFE. Foundations of Hyperbolic Manifolds.
150 EISENBUD. Commutative Algebra with a View Toward Algebraic Geometry.
151 SILVERMAN. Advanced Topics in the Arithmetic of Elliptic Curves.
152 ZIEGLER. Lectures on Polytopes.
153 FULTON. Algebraic Topology: A First Course.
154 BROWN/PEARCY. An Introduction to Analysis.
155 KASSEL. Quantum Groups.
156 KECHRIS. Classical Descriptive Set Theory.
157 MALLIAVIN. Integration and Probability.
158 ROMAN. Field Theory.
159 CONWAY. Functions of One Complex Variable II.
160 LANG. Differential and Riemannian Manifolds.
161 BORWEIN/ERDÉLYI. Polynomials and Polynomial Inequalities.
162 ALPERIN/BELL. Groups and Representations.
163 DIXON/MORTIMER. Permutation Groups.
164 NATHANSON. Additive Number Theory: The Classical Bases.
165 NATHANSON. Additive Number Theory: Inverse Problems and the Geometry of Sumsets.
166 SHARPE. Differential Geometry: Cartan's Generalization of Klein's Erlangen Program.
167 MORANDI. Field and Galois Theory.
168 EWALD. Combinatorial Convexity and Algebraic Geometry.
169 BHATIA. Matrix Analysis.
170 BREDON. Sheaf Theory. 2nd ed.
171 PETERSEN. Riemannian Geometry.
172 REMMERT. Classical Topics in Complex Function Theory.
173 DIESTEL. Graph Theory. 2nd ed.
174 BRIDGES. Foundations of Real and Abstract Analysis.
175 LICKORISH. An Introduction to Knot Theory.
176 LEE. Riemannian Manifolds.
177 NEWMAN. Analytic Number Theory.
178 CLARKE/LEDYAEV/STERN/WOLENSKI. Nonsmooth Analysis and Control Theory.
179 DOUGLAS. Banach Algebra Techniques in Operator Theory. 2nd ed.
180 SRIVASTAVA. A Course on Borel Sets.
181 KRESS. Numerical Analysis.
182 WALTER. Ordinary Differential Equations.
183 MEGGINSON. An Introduction to Banach Space Theory.
184 BOLLOBAS. Modern Graph Theory.
185 COX/LITTLE/O'SHEA. Using Algebraic Geometry.
186 RAMAKRISHNAN/VALENZA. Fourier Analysis on Number Fields.

187 HARRIS/MORRISON. Moduli of Curves.
188 GOLDBLATT. Lectures on the Hyperreals: An Introduction to Nonstandard Analysis.
189 LAM. Lectures on Modules and Rings.
190 ESMONDE/MURTY. Problems in Algebraic Number Theory.
191 LANG. Fundamentals of Differential Geometry.
192 HIRSCH/LACOMBE. Elements of Functional Analysis.
193 COHEN. Advanced Topics in Computational Number Theory.
194 ENGEL/NAGEL. One-Parameter Semigroups for Linear Evolution Equations.
195 NATHANSON. Elementary Methods in Number Theory.
196 OSBORNE. Basic Homological Algebra.
197 EISENBUD/HARRIS. The Geometry of Schemes.
198 ROBERT. A Course in p-adic Analysis.
199 HEDENMALM/KORENBLUM/ZHU. Theory of Bergman Spaces.
200 BAO/CHERN/SHEN. An Introduction to Riemann–Finsler Geometry.
201 HINDRY/SILVERMAN. Diophantine Geometry: An Introduction.
202 LEE. Introduction to Topological Manifolds.
203 SAGAN. The Symmetric Group: Representations, Combinatorial Algorithms, and Symmetric Functions.
204 ESCOFIER. Galois Theory.
205 FÉLIX/HALPERIN/THOMAS. Rational Homotopy Theory. 2nd ed.
206 MURTY. Problems in Analytic Number Theory.
Readings in Mathematics
207 GODSIL/ROYLE. Algebraic Graph Theory.
208 CHENEY. Analysis for Applied Mathematics.
209 ARVESON. A Short Course on Spectral Theory.
210 ROSEN. Number Theory in Function Fields.
211 LANG. Algebra. Revised 3rd ed.
212 MATOUŠEK. Lectures on Discrete Geometry.
213 FRITZSCHE/GRAUERT. From Holomorphic Functions to Complex Manifolds.
214 JOST. Partial Differential Equations.
215 GOLDSCHMIDT. Algebraic Functions and Projective Curves.
216 D. SERRE. Matrices: Theory and Applications.
217 MARKER. Model Theory: An Introduction.
218 LEE. Introduction to Smooth Manifolds.
219 MACLACHLAN/REID. The Arithmetic of Hyperbolic 3-Manifolds.
220 NESTRUEV. Smooth Manifolds and Observables.